The Complete Guide to
RV Electrical, Computer, Solar and Communications Systems
Working and Living Independently on the Road

William C. Meyer

© 2009 William Meyer
All Rights Reserved.

No part of this publication may be reproduced, stored in a retrieval system, or transmitted, in any form or by any means, electronic, mechanical, photocopying, recording, or otherwise, without the written permission of the author.

First published by Dog Ear Publishing
4010 W. 86th Street, Ste H
Indianapolis, IN 46268
www.dogearpublishing.net

ISBN: 978-160844-040-5

This book is printed on acid-free paper.

Printed in the United States of America

Table of Contents

About the Author v

Preface vii

Chapter 1 1
Planning
    Define Your Needs · RV Plans and Schematics · Current RV Facilities (Electrical, Cable) · Egress and Ingress Points · Weight Capacities

Chapter 2 7
Computer Equipment
    Computers · Monitors · Keyboard · Mouse · USB Connections · Wireless Networking/Internet Access · Printing/Scanning · The Paperless Office · Computer Backups

Chapter 3 66
The Electrical System
    The 120-Volt AC System · The 12-Volt DC System · System Monitoring · Maintenance · Case-In-Point Example

Chapter 4 98
Solar Power
    Is Solar for You? · System Components · Wiring Diagrams · Maintenance

Chapter 5 115
Communications
    Phone · Internet · Email

Chapter 6 186
An Illustrative Example

Appendix A 202
Vendor Reference List

Appendix B 204
Definitions

Appendix C 212
Helpful Links

Index 213

About the Author

William Meyer graduated from the University of Wisconsin–Madison with a degree in Mechanical Engineering in 1981. What followed was a successful 22 year career in the computer field, working for manufacturing companies and in the financial services business. Bill did this by taking a keen interest in the emergence of the personal computing era. Starting with the writing of computer software programs to get useful work accomplished with personal computers, Bill moved on to studying the art of fixing PCs and networking. Bill then got his Microsoft® MPC certification, which established his credentials in the computer field.

Bill is now president of his own business: the Kettle Moraine Computer Group. Continuous training and study in such topics as the Internet and e-commerce, along with the new wireless technology, keeps Bill and the company on pace to offer solutions that are timely and useful to business owners. Bill specializes in developing programs for heat transfer companies and adding e-commerce to existing web sites.

Bill is also a devoted RV enthusiast, who enjoys both working and relaxing while touring the United States. These days you can find Bill traveling the cities and countryside visiting clients, writing computer code, fixing PCs and working on his next RV or book project.

Do you have comments or questions regarding the content covered in this book? Please feel free to drop the author an e-mail at **william@williamcmeyer.com**.

Preface

The main purpose of this book is to act as a how-to reference manual for outfitting your RV or mobile home with the tools to support your working and life style and is written from the perspective of traveling full-time on the road. A secondary purpose is to supply information for persons to boondock where they wish and still have the electrical conveniences without having shore power. You will find topics on electrical systems, computers, solar energy and communications. Most all of the components referred to in the book are easily found at any computer retail store, hardware store and of course, on-line. You can save hundreds of dollars by installing your RV systems and have the pleasure and satisfaction of doing it yourself. The examples illustrated are a starting point to give you ideas on how to make changes to your RV.

You will find detailed pictures, schematics and illustrations throughout the book to guide you along the way. There is also a reference section for finding the components you need and to further explore a topic. Included is a glossary of technical definitions for terms and abbreviations. For the more technically inclined, there are sections interspersed all throughout the chapters that provide ancillary information on the topic being discussed. Go to these Sidebars to learn more about the technology or feature.

For those that work on the road, replicating your home or work office into a mobile office in the past has been difficult and frustrating to achieve because you could not accomplish all the same features and functions that are required to perform your job. As time has progressed, technology and people's inventiveness have continued to progress and seems to be progressing in an increasing geometric fashion. Obviously, this is good news for the mobile worker or mobile entrepreneur. Today, as of the writing of this book, the technology is available to match the performance and functionally of a brick and mortar establishment.

Look for these callouts to guide you through this book:

> **Sidebar**
>
> The sidebars interspersed throughout the chapters provide ancillary information on the topic being discussed. Go to the sidebars to learn more about the technology or feature.

In compiling this book, I have drawn from my background as a mechanical engineer and certified computer professional to develop this guide. My personal experiences as a full-time RV traveler and those of my colleagues and associates over the years have helped me put this reference together. Perhaps you will follow the ideas expressed in the manner that I have, but please don't feel key-holed into thinking that these are the only way to accomplish your goal. The examples illustrated are a starting point to give you ideas. Certainly, technology will continue to advance and simpler, more compact and efficient components and methods will evolve.

Who Should Read This Book?

You don't need to be an engineer or a computer scientist to read this book. The book is written at a beginning to moderate level so that most everyone will be able to follow it. There are also the sidebars that go into advanced detail for those that want to know the "nuts and bolts" about the topic. There are lots of examples and step-by-step instructions. The book assumes some familiarity with computers and the Windows® operating system (the examples in the book were done in Windows XP Professional). If you need instruction, web sites like Video Professor™ (www.VideoProfessor.com) can assist you.

Having some Do-It-Yourself experience is beneficial in the electrical and solar sections, but not mandatory. All electrical items are mapped out in full detail. For the most part you only need to know positive from negative.

The book covers a spectrum of topics that will allow you to construct a state-of-the-art mobile office or an efficient boondocking. The topics can also be taken individually if you are only interested in how to install solar power, for example.

Books of this nature use a lot of technical jargon and abbreviations, so you will find detailed definitions in Appendix B in the back of this book.

After reading the book, you should feel comfortable with each topic and there are plenty of references to point you to locations for finding additional information.

Safety Warnings

The contents of this book describe modifications, changes and additions to your Recreational Vehicle's electrical, mechanical and structural design. Please use caution when working on your vehicle. The components referred to in this book come with their own explicit safety and warning guidelines that we highly recommend that you follow.

The ideas and methods described in each chapter can easily be accomplished by the average Do-It-Yourselfer. If you feel uncomfortable or unsure about any procedure, there is a section in the back of the book under Vendor Reference List that has helpful links to assist you in accomplishing the task either by yourself or by consulting or hiring a trained technician.

Chapter 1 – Planning

Planning is an essential element for a successful and less frustrating endeavor. The time you put into researching and investigation will be worth the effort. You will also begin to get a feeling for the amount of time the project will entail and the associated costs. Most of the tasks in the planning stage are just collecting information and formulating ideas. It is too early at this point to get into design specifics.

DEFINE YOUR NEEDS

Start by creating a list of the items you need. Be sure to think in terms of needs versus wants and make separate lists for both. Your needs will be mandatory but your wants can be something to be added at a later date as your needs change or as a way to control the overall cost. At this point, don't worry about the brands or models of products you'll buy, this is just a generic, high level list of the items needed. Determining the specifics will come later as you shop for each item. Here's an example:

Need
- Laptop
- LCD monitor
- Printer/Scanner
- Telephone connection or cell phone
- Television
- Extra work lights

Want
- Fax capability
- Two LCD monitors
- External hard drive backup storage for laptop
- Solar power
- CB radio
- GPS Receiver
- DVD/Surround Sound

Creating lists as above will start to get you thinking of your overall power requirement (discussed in chapter 3). Keep in mind what level of autonomy from hookups (electrical, phone, television, etc.) are desired in the overall scheme of things. These will be important questions in the upcoming chapters.

RV PLANS AND SCHEMATICS

In the planning stage, you should try to obtain the floor plans for your RV and any schematic drawings from the manufacturer. The most helpful schematic will be of the electrical system. See Figure 2.1 and 2.2 for an example of a floor plan that also includes the electrical schematic for the RV. This information should be part of the operations manual for your RV or you may find it on the manufacturer's web site.

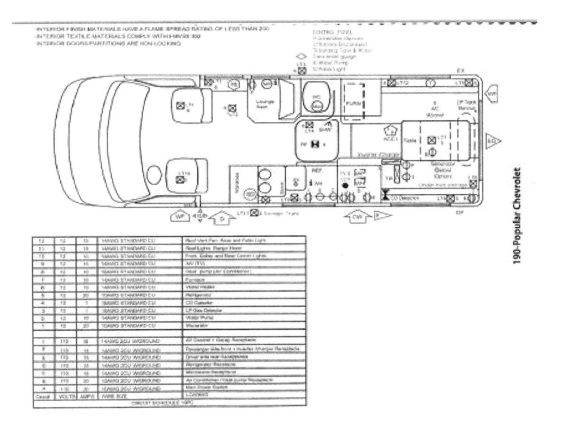

Figure 2.1 Floor plan and electrical schematic for Home & Park 190-Popular.
(Copyright © Hanmar Motor Corporation)

Figure 2.2 Legend for electrical schematic used in the Home & Park 190-Popular.
(Copyright © Hanmar Motor Corporation)

The electrical portion of the schematic is probably the most useful and will come into play in chapter 3 when power requirements and locations will be determined. For now I would suggest making copies of the documentation and starting file folders to collect the information in one location for future reference. It will also be handy as you pick out components to collect the literature for each and keep them in your project folder. I often jot notes on them to include things like purchase source, purchase price and electrical details (i.e. power usage and voltage) if applicable.

CURRENT RV FACILITIES (ELECTRICAL, CABLE)

With the aid of the schematics and an inventory of the RV electrical system, you should be able to determine the total electrical capacity of the AC and DC branch circuits. This will be needed in chapter 3 to access excess electrical capacity already installed. See Figure 2.3 for an example of a typical RV fuse panel.

Fuse Panel

Fuse Panel Cover

Fig2.3 Typical RV fuse panel showing the AC circuit breakers and the DC plug-in fuses.

You should also take note of the electrical cable wire gauges at this point if possible. This includes the branch circuits in the interior of the cabin and wire gauges used in between the inverter and house batteries (see Figure 2.4). This information will be used if additional load is added to an existing circuit to be sure the wiring is adequate.

There is most likely coaxial cable pulled through the RV for a television antenna, or perhaps a satellite dish. If so, make a note of it at this time for it may be piggybacked for other uses.

Figure 2.4 Electrical cables used between the inverter and the house batteries. Wire gauge can be found on the plastic casing of the cable or can be approximated by the diameter of the cable.

EGRESS AND INGRESS POINTS

Depending on your needs list, there may be a number of items that need to be connected between the outside of the RV and the inside. Unfortunately, this is one of the more difficult tasks associated with modifying an RV after it leaves the factory. But, it is not impossible to facilitate these connections, it just takes extra planning and consideration. Most recreational vehicles and motor homes these days are made of a wood/fiberglass composite to keep weight to a minimum. The good thing is that it is easy to drill through this exterior surface. The tricky part is making sure you do not drill through an object that may be located just below the surface (say between the roof and the fiberglass insulation and decorative vinyl padding. Based on the schematics you obtained, you should be able to tell if you are close to such an object. If in doubt, resources such as the dealer where you purchased the RV, the manufacturer, or an acquaintance that owns a similar rig might be able to assist you.

There are also intrinsic egress and ingress points that are already part of the existing RV. Examples include:

- The toilet roof vent
- The roof vent fan
- Existing cables routed outside the cabin
- The refrigerator vent
- The cook top exhaust vent

Again, at this point, it is just something to keep in the back of your mind as you make plans.

WEIGHT CAPACITIES

As shipped from the factory or dealer, your RV has an estimated gross vehicle weight and an associated maximum allowed weight. It is highly unlikely that you will come close to exceeding the maximum by adding the type of items we are discussing here, but for safety's sake, weight should be taken into consideration if you think that you are approaching the limit. See Figure 2.5.

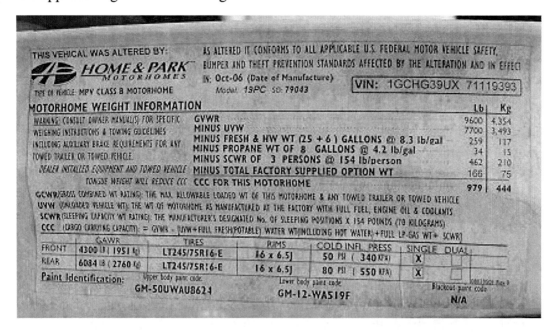

Figure 2.5 Example weight label typically located on the drivers side door.

Chapter 2 – Computer Equipment

These days, it's hard to imagine not having access to a computer for extended periods of time. Casual as well as business computer users take for granted that a computer will be within easy reach and that there will be an Internet connection available. This trend is apparent for full-time and part-time RV enthusiasts just by the amount of advertising camp sites and resorts do that sport "Free Wi-Fi Internet Access!"

This chapter goes into detail about the type of computers geared for mobile use, the support equipment required and some ideas that will make your computer experience more useful and enjoyable.

COMPUTERS

Computers come in so many assortments, colors and shapes, that even seasoned computer users find it confusing. An entire book could be dedicated to choosing the "right" PC, so it's beyond the scope of this book to cover all the varieties.

Regardless of the hype salespeople will present to you, there really are only a few basic features you need to keep in mind when purchasing (or upgrading) a PC for your RV. First, it is assumed that the computer of choice for the RV will be a laptop. Laptops are as powerful as their desktop cousins, and since space and weight are important factors, it is the recommended way to go. Also, as a sidenote, since computers continue to improve in performance and capacity, the information presented here is current as of the publishing of this book.

Here are the recommended features:

- 1.5 GHz processor or greater
- 512 MB RAM or greater
- 40 GB Hard Drive or greater
- Windows XP Professional or Windows Vista (Home Premium or Ultimate Edition)

Other specifications like the following are purely a matter of individual preference:

- Brand
- Screen Size

- Weight
- Optical Drive (CD-R, DVD-R, DVD-RW, etc.)
- Battery Life

The next topic, monitors, might seem to be out of place since we were limiting the discussion to laptop computers which have their own built-in display, but many computer users, and especially business users, may want the option of having multiple monitors.

MONITORS

If you have the space and a desire for more viewing freedom with your laptop computer, read on. This section discusses the laptop's built-in monitor port for adding an extra monitor. It will be assumed that the extra monitor would be an LCD type, but CRT monitors also work, even though they are very bulky.

After you've used a computer that has multiple monitors, you'll wonder how you ever got along without it! See Figure 2.6 which shows multiple programs open at once. With one monitor, you usually only view one program at a time while the others are either in the background or minimized. With multiple monitors, you can surf the web while still seeing your email program and working on a Microsoft® Word document.

Figure 2.6 Multiple LCD monitor setup.

Your laptop will already have the capability to have two monitors, since laptops come with an extra monitor port. To have more than two monitors, an additional video card is required for each additional monitor. Products like USB Gear's USB 2.0 Video Card Adapter (see Figure 2.7) plugs into a USB port on your computer so you can plug in another LCD display.

Figure 2.7 USB 2.0 Hi-Speed USB Video Card Adapter SVGA.
(Courtesy of USB Gear – http://www.usbgear.com)

Microsoft® Windows Multiple Monitor Support

Windows makes it possible for you to increase your work productivity by expanding the size of your desktop. Connecting up to ten individual monitors, you can create a desktop large enough to hold numerous programs and windows.

You can easily work on more than one task at a time by moving items from one monitor to another or stretching them across numerous monitors. Edit images or text on one monitor while viewing web activity on another. Or you can open multiple pages of a single, long document and drag them across several monitors to easily view the layout of text and graphics. You could also stretch a Microsoft Excel spreadsheet across two monitors so you can view numerous columns without scrolling.

One monitor serves as the primary display and will hold the logon dialog box when you start your computer. In addition, most programs will display windows on the primary monitor when you initially open them. Different screen resolutions and different color quality settings can be selected for each monitor. Multiple monitors can be connected to individual graphics adapters or to a single adapter that supports multiple outputs.

Use Display in Control Panel to configure the settings for multiple monitors.

KEYBOARDS/MICE

With laptop computers, the keyboard and mouse are built in as you would expect. The keyboard is not full size and the mouse is operated via a touch pad with your finger. Some people have issues with the smaller keyboard and more than a majority of people don't like the control of a touch pad mouse.

There are alternatives, though (see Figure 2.8). A regular button mouse with a scroll wheel can be purchased with a USB connection to plug into the laptop. This gives you the control you're used to with a desktop computer. Mice also come in a wireless version if you don't like the cords on your work surface. There are also USB versions of keyboards that will connect to your laptop. Wireless keyboards are also available. See Vendor Reference List.

Figure 2.8 Wireless keyboard and mouse. In the confines of limited space of an RV, these options are worth the extra cost.

USB CONNECTIONS

With the discussion we've had regarding all of the USB type accessories, the question arises as to where to plug in each device. Typically, a laptop computer will have two or three USB ports (see Figure 2.9). When you have exceeded this limit, a USB hub (see Figure 2.10) can be added to the laptop to increase the number of available ports (Typical hubs come in 2, 4 and 8-port configurations).

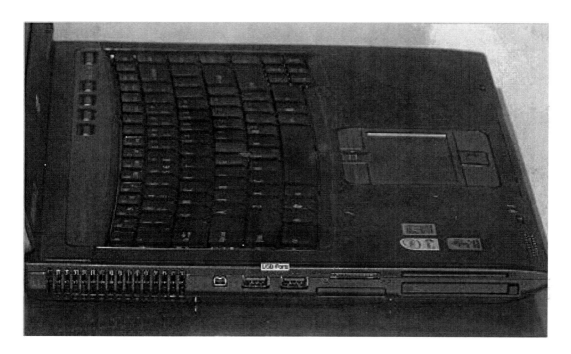

Figure 2.9 USB ports in a laptop computer.

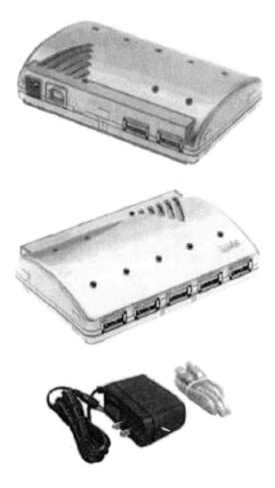

Figure 2.10 USB hub for plugging in additional devices.
(Courtesy of USB Gear – http://www.USBGear.com)

You will probably find that a USB hub is going to be mandatory when you consider all the devices that are now connected via USB ports (printer, scanner, wireless mouse, digital camera, etc.). See the sidebar below for more information about choosing USB hubs.

USB hubs

A USB (universal serial bus) hub is a device that allows many USB devices to be connected to a single USB port on your computer or to another hub.

Desktop and laptop computers usually come with 2 or 3 USB hubs built in. Separate USB hubs come in a wide variety of different shapes and sizes that will have multiple USB connections such as 4, 6, 8, etc., (referred to as ports). These hubs come in designs that can be directly connected to the computer without a connecting cable, or ones that are intended to be connected with a special USB cable. When purchasing cables, look for the product to be labeled as a "USB A-B Cable." Typical cables are sold in 6 ft or 10 ft lengths, and by inherent electrical design are limited to 16 ft (5 Meters). A hub can be used as an active USB repeater to extend cable length for up to 5 lengths, or 80 ft.

USB hubs are powered in one of two ways. A bus-powered hub is a hub that draws all its power from your computer's USB interface. It does not need a separate power connection. However, many devices require more power than this method can provide, and will not work in this type of hub. (The product literature for the device will indicate if a powered hub is required. If in doubt, you can always use a powered hub for any USB device without issue.) In contrast, a self-powered hub is one that takes its power from an external power supply and can therefore provide full power to every port. Many hubs can operate as either bus-powered or self-powered hubs.

Finally, USB hubs are currently designed to two specifications known as the USB 1.1 and USB 2.0 Specification which affect the data transmission speed of the hub. Older hubs designed to the USB 1.1 spec. are rated at 12 Mbps (12 million bits per second) while the USB 2.0 spec is rated at 480 Mbps (480 million bits per second). It is important to know that in common language (and often product marketing) USB 2.0 is used synonymously with "high-speed." Since the USB 2.0 specification, which introduced high-speed, incorporates and supersedes the USB 1.1 specification, any compliant full-speed or low-speed device is still a USB 2.0 device. If a full-speed device is plugged into a hub (or a port on the computer) that is designed as a 1.1 specification, the device will operate, but at the lower speed. Therefore, it you want high-speed, both the computer and hub must be designated as USB 2.0.

WIRELESS NETWORKING/INTERNET ACCESS

There is a high probability that your laptop already has built-in wireless networking, since the majority of newer laptops include it as a standard feature. Although the wireless feature can be used to connect you to other laptops and PCs on your network, we will be limiting the discussion to how wireless applies to having Internet access.

For the most part, wireless networking to an Internet connection (a.k.a. Hotspot or Wi-Fi) is automatic when your laptop is in range of the Wireless Access Point (WAP). If Internet access is free and no access code is required, you'll connect automatically. If not, the campsite or location you're at will provide an access code for you to connect.

Sometimes connecting to the Internet does not go smoothly and becomes a frustrating endeavor. I've included step-by-step instructions starting on the next page to get you connected on-line. Please note that these instructions apply to the Microsoft© Windows wireless connection manager (known as the Windows Zero Configuration (WZC) Service). If you have enabled another program to manage the wireless connection (which may have been installed with your wireless networking software), use that software and follow its directions.

The first step is to verify if Windows Wireless Networking is turned on.

<u>Verifying Windows Wireless Networking</u>
1. Open Windows Control Panel
2. Double click on **Network Connections**.

[INSERT NetworkConnections.jpg]

3. The **Network Connections** window will open and there will be a group called **LAN or High-Speed Internet.** From the list, click on **Wireless Network Connection**. Under **Network Tasks**, a link will become visible labeled **View available wireless networks**. Click on the link.

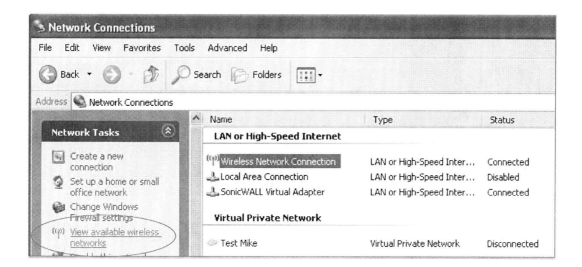

If the following message appears in the Wireless Network Connection dialog box, you will need to proceed to the **Configuring Windows Wireless Networking** section below.

18   The Complete Guide to RV Electrical, Computer, Solar and Communications Systems

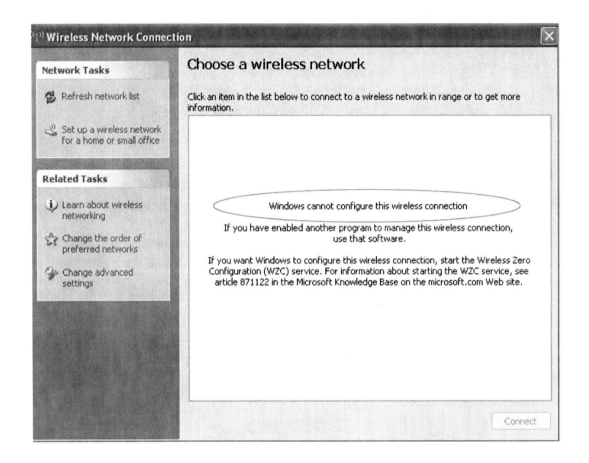

Configuring Windows Wireless Networking

    1.   Open Windows Control Panel
    2.   Double click **Administrative Tools**

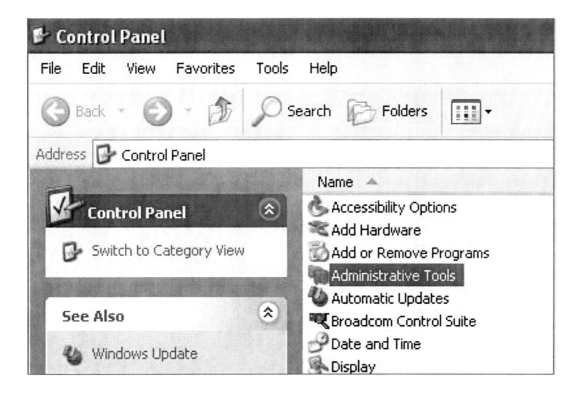

3. Select **Services**

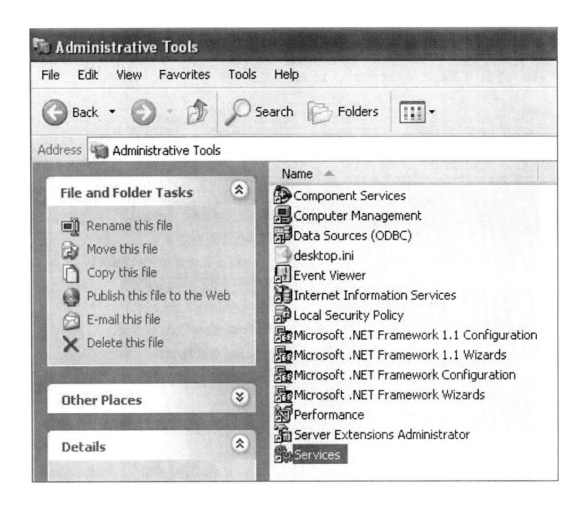

From the **Services** list, scroll down the list and find **Windows Zero Configuration.** Click on **Windows Zero Configuration** to select it, then click Start **the service** on the left hand side. In a moment, the Status column will indicate that the service has started. You are now ready to view available wireless networks. Continue below with **Connecting to a Wireless Network.**

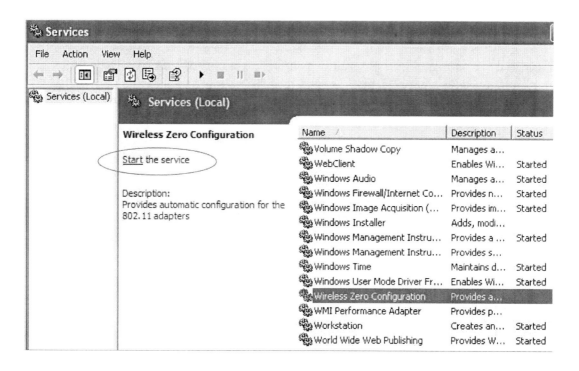

22 The Complete Guide to RV Electrical, Computer, Solar and Communications Systems

Connecting to a Wireless Network

Now that the configuration program is running (which it will do from now on), you may find a wireless network and connect to it. To do so:

1. Go back to the Control Panel
2. Double click on **Network Connections**.

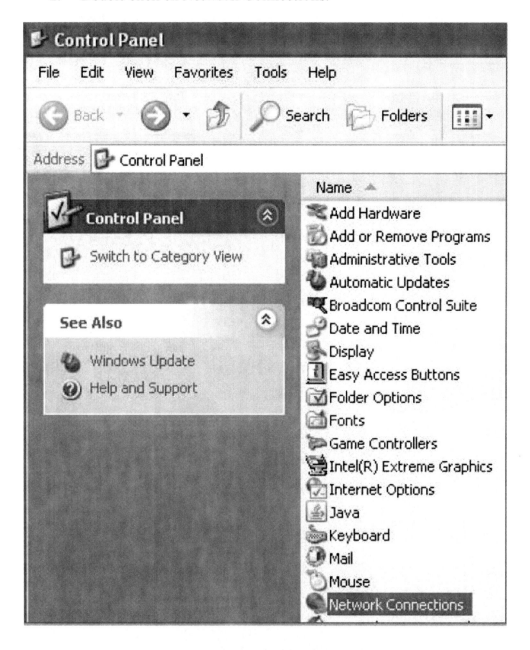

3. The **Network Connections** window will open and there will be a group called **LAN or High-Speed Internet.** From the list, click on **Wireless Network Connection**. Under **Network Tasks**, a link will become visible labeled **View available wireless networks**. Click on the link.

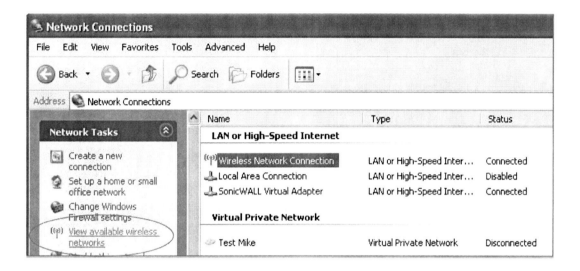

4. The following dialog box will appear asking you to choose a wireless network. All networks within range will show up on the list. The name of the network is beside the antenna (in this case **linksys**). In this example, the network requires a key for connection. This is indicated by the lock symbol below the name. The key will be something you obtain from the Wi-Fi manager. (If it is an open site, click on **Connect** at the bottom.)

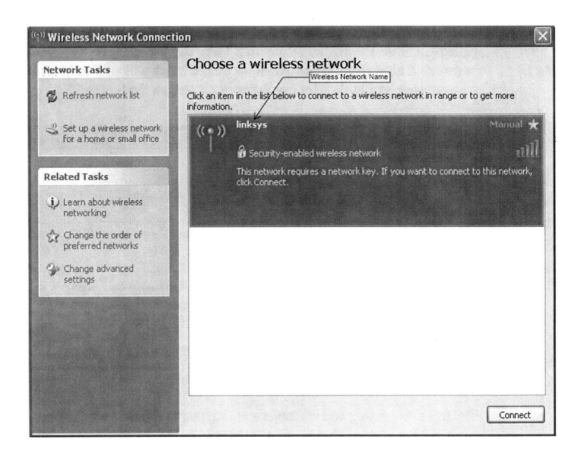

When connecting to a secure site (locked), you will be presented with the following dialog boxes after you click on the **Connect** button. Type in the key, confirm the key and hit Connect. You will be connected after a brief pause and the connection status indicated next to the yellow star in the upper right-hand corner will say Connected.

Wireless Network Connection

The network 'linksys' requires a network key (also called a WEP key or WPA key). A network key helps prevent unknown intruders from connecting to this network.

Type the key, and then click Connect.

Network key:

Confirm network key:

[Connect] [Cancel]

Paid vs. free internet access

The terms "**Hotspot**" and "**Wi-Fi**" (acronym for wireless fidelity) are becoming common terms in the RV traveler's vocabulary. In general it means a venue that offers Wi-Fi access to the public on a paid or free basis. For venues that have broadband service, offering wireless access is as simple as purchasing one Access Point (Wi-Fi router) and connecting the Access Point with the gateway box. Hotspots are often found at restaurants, train stations, airports, libraries, hotels, hospitals, coffee shops, bookstores, fuel stations, department stores, supermarkets and other public places. Many universities and schools have wireless networks on their campuses.

Commercial hotspots provide a captive portal (gateway) that users are redirected to for authentication and payment to obtain access to the provider's Internet connection. Payment options using credit cards, PayPal, BOZII, iPass, or other payment services are presented when you connect to the Wi-Fi site. Typically you can purchase daily, monthly or annual plans.

Free hotspots are readily available across the country. Open public networks are the easiest way to create a free hotspot. All that is needed is a Wi-Fi router. To find where the free spots are located, visit sites like www.jiwire.com or do an Internet search for additional listings.

Technical details on making connections to the Internet

If you're wondering how your computer actually gets connected to the Internet at a hotspot location, read on.

All Wi-Fi providers basically use the same components for getting their users attached to the Internet. This includes the use of special wireless equipment that conforms to the IEEE 802.11b/g wireless specifications to manage the process as shown below:

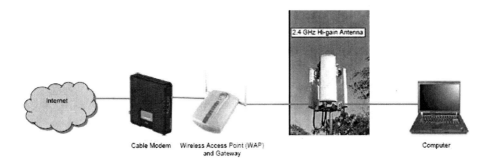

The first component in the link is the DSL (Digital Subscriber Link) cable modem that connects to the telecommunication service provider's network. This device provides the interface to the nationwide network (Internet).

Connected to the modem is the Wireless Access Point (WAP) that also does the routing between the hotspot's network and the public network. (This is also referred to as the gateway software.) The WAP can act as a Firewall, which is a combination of hardware and software that provides a security system to prevent unauthorized access from the outside public network to the internal hotspot network. The firewall prevents direct communication between network and external computers by routing communication through this proxy device outside of the network. The proxy determines whether it is safe to let a file pass through to the network.

Next, a high gain antenna is employed to broadcast the Wi-Fi signal to your computer. The size of the antenna is chosen based on the intended coverage area. In large parks there may also be "repeater" Access Points positioned throughout the park that extend the signal to points that are far from the WAP.

When your computer attempts to go onto the Internet, it is handed an IP (Internet Protocol) address from the WAP so the computer can be uniquely identified (think of it as a street address for your home where all addresses must be unique). In this fashion, the WAP is performing the function of a Dynamic Host Configuration Protocol (DHCP) server. This service provides dynamic IP addresses and distributes other configuration parameters (like the IP address of the gateway) to your computer. This prevents address conflicts and helps conserve the use of client IP addresses on the network.

Once the computer has its own IP address, the WAP can forward your requests to the Internet (e.g. *www.yahoo.com*, IP address *63.201.155.210*) and send responses back. In conjunction with DHCP, the WAP points to the Domain Name System (DNS) servers on the Internet. DNS associates various information with domain names, most importantly, it serves as the "phone book" for the Internet by translating human-readable computer hostnames, e.g. *www.yahoo.com*, into IP addresses, e.g. *63.201.155.210*, which networking equipment needs to deliver information. The DNS is an essential component of the contemporary Internet.

Even though your laptop has built-in wireless, the following section is going to talk about the rational of adding a third-party external wireless card to your laptop. The logic in doing this will become apparent after you've pulled into your assigned site for the night, open your laptop to connect to "Free Internet Access" and find that no wireless networks appear on the list!

Although this will be a disappointment, there are a few reasons why this will happen. Wi-Fi is a low power radio signal that connects your laptop to the Wi-Fi host. There are many things that can get in the way and cut the signal. Objects like metal or living things like trees or bushes act as a short. This means the signal will go to the ground and not to your receiver. The real problem comes when people want to use Wi-Fi inside their RV. Things like metal, plastic, fiberglass, windows and dry wood may cut down the signal to the point that it cannot be picked up by the laptop. The farther away your laptop is from the Wireless Access Point, the weaker the signal. Campsites and resorts would find it cost prohibitive to provide a strong single to every square foot of property, so there are always going to be "weak-spots" that will have trouble receiving the signal.

There is a solution though.

You can extend the Wi-Fi range by adding an external antenna. This can be accomplished by purchasing a wireless network card that comes equipped with a connection for hooking up an antenna cable (see Figure 2.11). You then purchase an antenna and antenna cable so you can mount it on the roof of your RV. This will greatly enhance your ability to pick up the signal.

Figure 2.11 External wireless network card with a connection for an antenna. This example plugs into one of the laptop's USB ports. (Courtesy Fleeman Anderson & Bird http://www.fab-corp.com)

There are numerous antennas available at a variety of prices (see Vendor Reference List). Pricing varies based on the type of antenna and the amount of "Gain" the antenna produces (signal amplification measured in decibels (dB)). I recommend an omni-directional type antenna (see Figure 2.12 and Figure 2.13), so that you can pick up the signal from the Wireless Access Point no matter what direction you are facing. **(NOTE: Be sure to select a 2.4 GHz antenna. These are designed for wireless networking.)** When it comes to what gain level you should choose, buy as much as you can afford. A typical 15 dB gain antenna is under $100 and will increase your range from a quarter mile to a mile depending on conditions and other factors.

Figure 2.12 Omni-directional antenna for receiving wireless signals. (Courtesy Fleeman Anderson & Bird http://www.fab-corp.com)

Figure 2.13 Example of an Omni-directional antenna installed on an RV.

How do hotspots make long range outdoor and campus type Wi-Fi services available?

Let's look at an RV park as an example.

The first step is to have Broadband internet access to the location obtained from the local phone company, cable company or via satellite.

Once this is obtained, the second step is to develop a Site Plan that details the size of
the park (acreage), the layout (square or other), the location of the broadband service and the typical type of foliage (heavily wooded, sparse trees, etc.). Terrain and large structures are also taken into account. Foliage, terrain and buildings will have an effect on the signal reaching all areas of the park.

After this information is gathered, selection of the right type of equipment, which includes the Wireless Access Point, Repeaters (amplifier) with enough power to deliver the signal to the guests and the proper type of antenna(s). The critical point is that the signals are picked up at the locations in the park that they are needed ("can you hear me now?"). This is accomplished through the use of strategically placed antenna and signal repeater stations.

Wireless signal strength and connection speed.

If you're curious about how good your wireless connection is at any given time, Microsoft Windows has a utility that shows you the current status of your wireless connection. Open the Control Panel and choose Network Connections. The Network Connections window will open and there will be a group called LAN or High-Speed Internet. From the list, click on Wireless Network Connection. Under Network Tasks, a link will become visible labeled "View available wireless networks". Click on the link.

When the Network Connection Status window opens you'll see the signal strength that illustrates the strength as a series of bars (5 bars being the best) or it may say High, Medium or Low. The connection speed will vary based on the signal strength and the type of connection. The current wireless specifications available are IEEE 802.11b at 11 Mbps (million bits per second) and 802.11g at 54 Mbps transfer speed. Due out soon will be the 802.11n specification that will allow manufacturers to build wireless equipment capable of 100 Mbps. Therefore, depending on the wireless adapter in your computer and the wireless equipment used by the Wi-Fi provider, the maximum connection speed will be the slower of the two devices in communication with each other, so either 11 or 54 Mbps. The connection speed will decrease as the signal gets weaker.

PRINTING/SCANNING

Printing and scanning are a necessity that no office (mobile or otherwise) can live without. These devices have come down in size and in price, so they are not only affordable, but they can fit into an RV without taking up too much space.

Printing is probably going to be on your "needs" list although scanning may be a luxury that goes on the "wants" list. If you need both, there are many all-in-one printers that have both functions (along with copying, and faxing), but these machines are quite bulky and it may be hard to find space for them. Also, the faxing component will need a land line telephone connection. This, of course, can be obtained at most campgrounds and resorts, but will not be something you can do while travelling.

As an alternative to faxing, it is often simpler to scan in the document you would normally fax and email it to the recipient. If you decide to do this, you could do what I have done: purchase a separate printer and scanner which are easer to stow when not in use. See Figure 2.14. The printer resides in a location in the RV where I can always have it plugged in to the laptop. I keep it powered off when not in use to conserver electricity. The scanner gets tucked away in a cubbyhole and I bring it out only as needed, since I don't scan as frequently as I print. The scanner is also used to copy documents and to create PDF files. See An Illustrative Example.

Scanner

Printer

Figure 2.14 Color printer and scanner. (Courtesy Cannon Inc. http://www.canon.com and Hewlett-Packard http://www.hp.com)

Keep in mind that printers and scanners consume much more power than your laptop, even when they are just turned on and sitting idle. You will want to manage this power usage diligently if you are not connected to shore power (i.e., boondocking or dry camping). The best thing to do is to turn them on only when needed and immediately turn them off when done.

THE PAPERLESS OFFICE

Just about everyone will agree that there is no such thing as a paperless office. But when it comes to a mobile office, striving to become as paperless as possible goes a long way when you're in tight accommodations.

There is clearly limited room in your RV, so putting as many documents as possible into digital form and storing them on your laptop will save a lot of space and give you the ability to keep them close at hand. To digitize your documents you will need a scanner to get them into digital format. I take the approach of scanning the documents I need to my laptop, making backup copies to my external storage drive (see Figure 2.15) and throwing the original away. (If the original MUST be kept for whatever reason, I ship those documents to my relative who stores them in a filing cabinet set aside for my items.)

Figure 2.15 Example of a 256 GB external backup computer storage connected to laptop via USB cable. (Courtesy Western Digital http://www.westerndigital.com)

Here are some examples of what I scan to electronic format:

- Receipts
- Product manuals
- RV maintenance records
- Personal documents (e.g. Insurance, bank accounts, financial statements, etc.)
- Pet records
- Client correspondence
- Credit card records and correspondence
- Household and RV inventory records

The list could go on indefinitely it seems. Take a look inside your paper filing system and prune out what could be scanned to digital format.

I also put a number of items I get off the Internet into digital storage. For example, I have compiled a list of Wi-Fi hot spots off the web and downloaded the list (organized by location) to my laptop. I do this in case I want the information, but don't have access to the Internet when I need it. I do the same for data compiled by The Weather Channel (www.weather.com) for the monthly average temperatures and rainfall records for the various cities where I travel so that I can look up this information without an Internet connection. The same could be done with maps and information about campsites, resorts, state parks, etc.

Finally, I make use of a digital camera not only for taking personal photos of my travel experiences, but for also recording information regarding my RV inventory. I photograph my belongings like computers, monitors, printers, etc. I then edit the photo to include model numbers, serial numbers, and cost, and store this on my laptop. I also make a CD copy and send that to my insurance agent so they have a record of what I own in case of theft or loss. (I update it annually so that it is current and that I have enough replacement coverage.)

COMPUTER BACKUPS

Enough cannot be said about the importance of backing up the information on your laptop. This section will explain my philosophy and strategy for making backup copies of my important files.

First, there needs to be a backup device attached to the laptop. I personally use an external hard drive (Refer back to Figure 2.15) since I have a large quantity of files. You may find a 2-GB or 4-GB flash drive (a.k.a. Thumb drives, memory sticks. or pen drives) will hold all your data at less cost. See Figure 2.16. Also see sidebar below – How big of a backup drive do I need? Either way, they can be used as the storage medium for your valuable information.

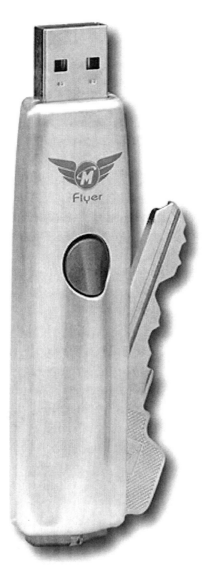

Figure 2.16 A 2 GB USB flash drive external computer storage device. (Courtesy Memorex http://www.memorex.com)

How big of an external disk drive do I need?

When making the decision about the type and size of an external disk drive to purchase, keep in mind that you don't need to back up everything on the computer. Employing the best practices used in the computer support industry, you will find below the strategy used to safeguard your information. Based on this, you will know which files need to be backed up and how much space is required on the external disk drive.

First, keep in mind that everything on the computer is stored in files on the hard drive. These files can be generally categorized as 1) Program files, 2) Data files and 3) System files.

Program files are the ones put on the computer during installation of the program, e.g. *Microsoft® Office*. Therefore, they are always "backed up" if you have the original program CDs and do not need to be backed up again. If your computer fails and the hard drive or entire computer needs to be replaced, you will restore the program by re-installing it from the CD. (This logic also applies to programs you may download off the Internet. If they need to be restored you would go back to the Web site and download them again.)

Data files are the most important files to backup. They are defined as the files you create when working with a program, e.g. *MyDocument.doc* produced with Microsoft Word. Keep track of where you create your files and back them up as often as necessary. (For example, in the production of creating the Microsoft Word files for this book, I backed them up everyday after I worked on them.)

System files contain, among other things, the Windows settings for your computer. Items like the background color and image for the display screen, icons on the desktop, etc. This information is included when you backup the files located in your **Documents and Settings** folder. (For example, on my computer these are backed up when I select the **C:\Documents and Settings\Bill** folder in the backup selections. The Documents and Settings folder contains subfolders for each user that logs into the computer and also contains the ***My Documents*** folder.) Windows and programs installed on your computer also keep track of their own system settings in the ***Windows Registry*** files. These are backed up when you select ***System State*** in the backup selections.

Now that you know which files to backup, determine how much space you'll need. Most backup programs will tally up and display the total file size of all the files and folders selected in terms of number of bytes, e.g. 1.3 GB. Use that number to judge the disk space needed to backup your computer and add some for future growth.

If your backup program does not show the total backup size, you can get the information by opening My Computer and navigating to the folder in question and getting it's properties. For example, navigate to c:\documents and settings. Right-click on the **Documents and Settings** folder name and choose **Properties** from the menu and under the **General** tab look for the item called **Size on disk** which will show the total bytes for that folder.

Once the backup drive is in place, you will need backup software to perform the backup. Some devices come with their own software which can be used to manually or automatically backup your files on a schedule you pick. If no software is provided, Windows comes with backup and restores software. I'll give examples of both below.

The external hard drive that I selected is a Western Digital© My Book™ (www.westerndigital.com) that plugs into my laptop with a USB connection. It comes with a program called WD Backup™ which can perform backups and restores. The example that follows shows how to pick the files for backup and how to schedule the backup to run once a week (for WD Backup version 1.0.2.16). (Remember that scheduled backups will require that the laptop and My Book™ be turned on at the time of backup, so plan accordingly. If the devices are not on, the backup will fail and be rescheduled for the following week.)

Backing Up Files with the WD Backup Software

1. Start the WD software which is found under Start -> All Programs -> WD Backup -> WD Backup. On the main menu choose Back up.

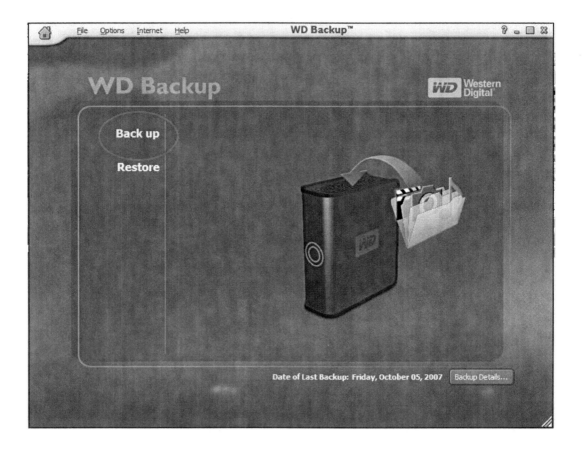

2. Select **Schedule Backup.**

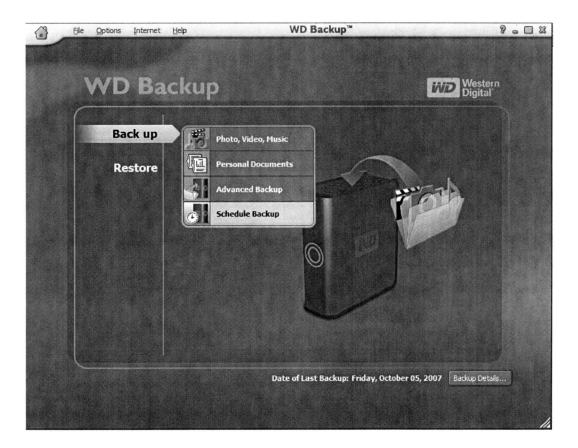

3. Pick **Advanced Backup** from the Backup Choices

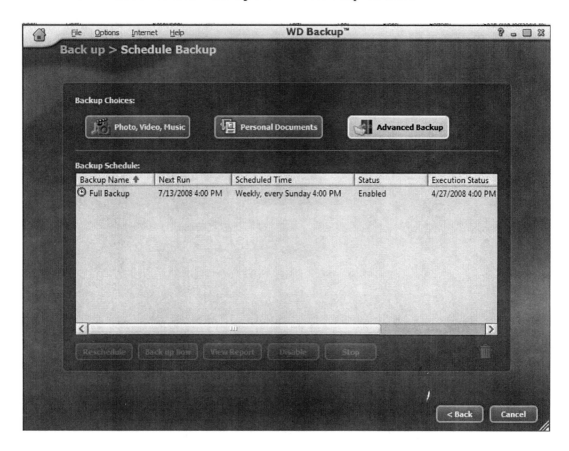

4. Select the files to backup by expanding the list in the left-hand pane and putting a check next to the folder to backup. Individual files can be selected or unselected by checking or un-checking the file names in the right-hand pane. (Leave **Full Data Backup (Include all selected files)** selected under item **#2 What kind of backup would you like to create?**) Click **Next** when done.

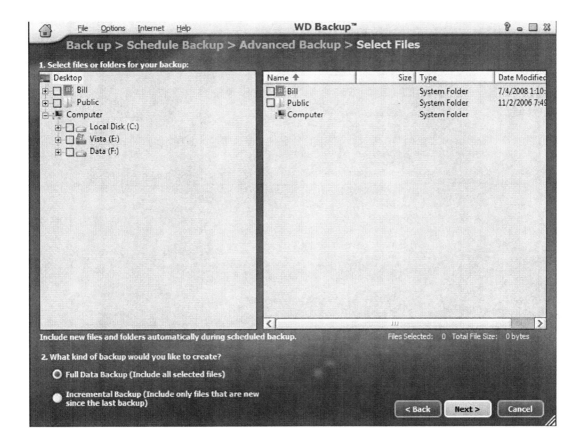

5. Finally, give the backup a name in the box provided and add any comments. The **Backup Destination** will default to the laptop's hard drive. You WILL need to change the location to the external drive by clicking on the **Browse** button. Navigate to My Computer, the select the device labeled **My Book (G:)**. (Note: The drive letter may be different for your computer.) Click **OK** to continue. Then, under **Schedule Settings,** pick a frequency, date and time to run the backup. When ready, click on the **Schedule It** button. You will then see verification that the job is scheduled on the next screen.

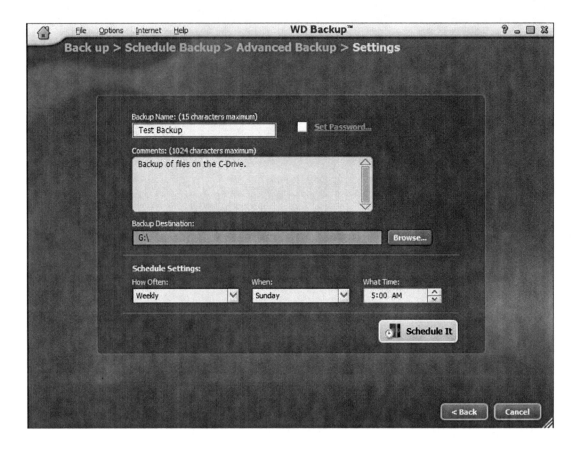

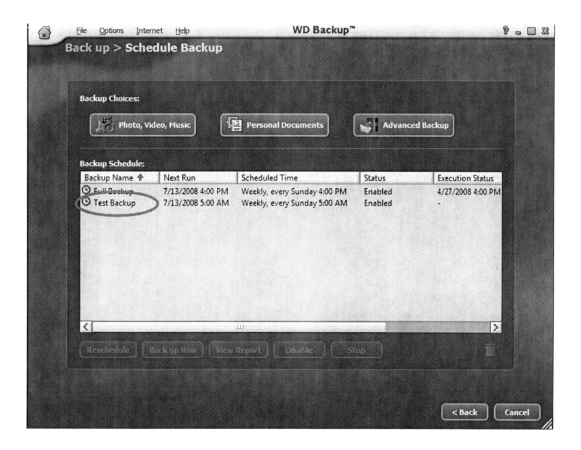

Once you create the backup job, you can run it manually anytime by opening WD Backup -> Back up -> Schedule Backup. The job will be listed under **Backup Schedule.** Click on the job and then click on the **Back up Now** button. Make sure the destination is the external drive, then click the **Back up** button to start the job.

Restoring Files with the WD Backup Software

You will at some point need to restore files. Here are the steps:

1. Open the WD backup program and choose Restore.

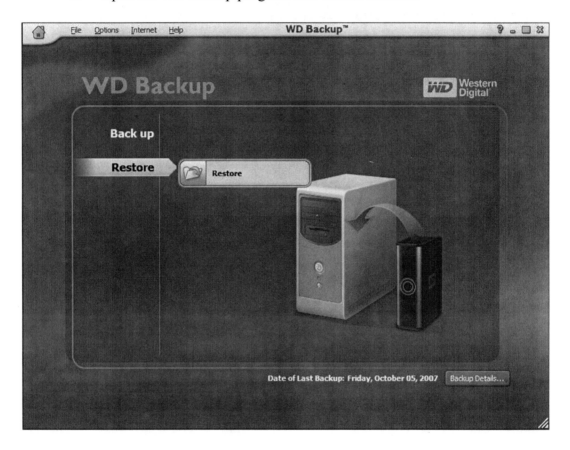

2. Select the Backup to restore from by clicking on the backup name (There will be a list of all backups on file by date.) and then clicking on the **Open** button and the **Next** button on the following screen.

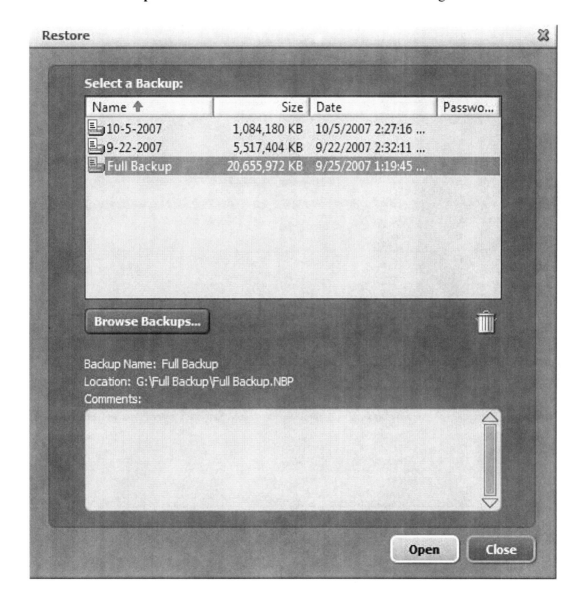

3. Select **Advanced Restore – Select files individually.** Click **Next.**

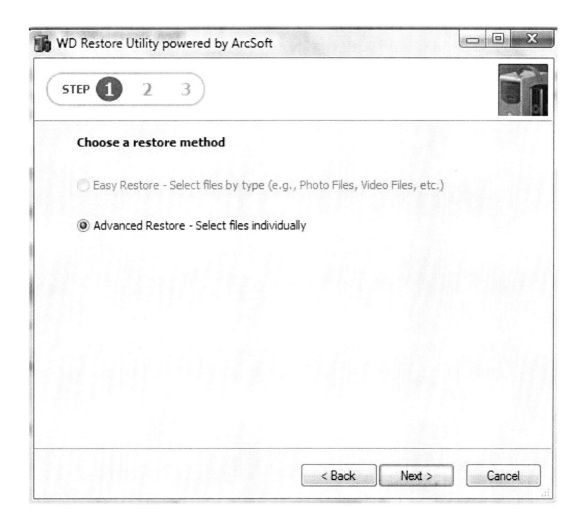

4. From the left-hand pane, select the folders to restore by putting a check mark in the box next to the folder name. (You can navigate down the folder list to select individual files in the right-hand pane. Anything with a check mark next to the name will be restored.) Click **Next**.

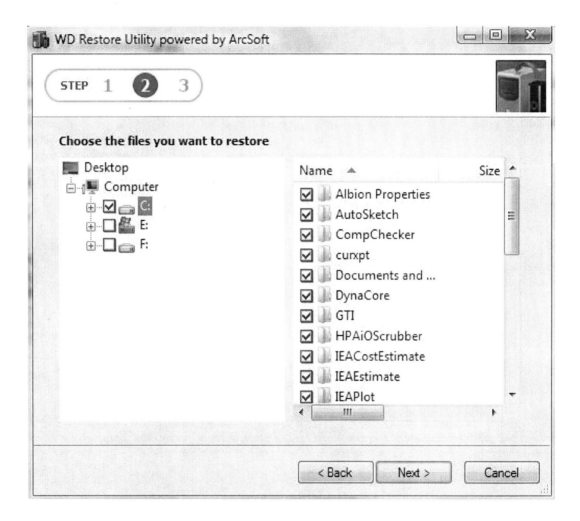

5. Next, choose the restore destination. The usual selection is **Restore the selected files to their original locations,** but you could redirect the restore to another location by selecting **Restore the selected files to one specific location.** Click on the **Restore** button to start the restore. (Note: If the files already exist, you will asked to confirm that you want to overwrite them.)

6. You will receive a **Restore Report** when it is finished. Click **OK**, then **Finish**.

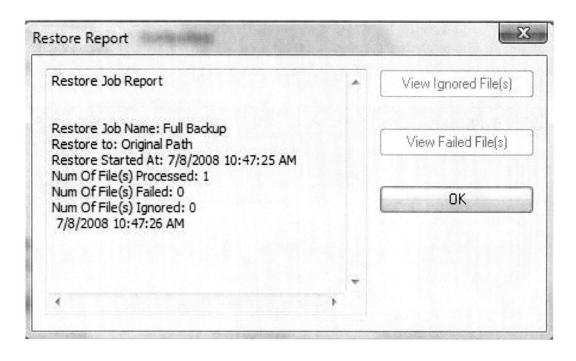

The alternate method is to use The Windows Backup program.

Backing Up Files with the Windows Backup Software

1. Open the backup program by going to Start -> All Programs -> Accessories -> System Tools -> Backup. The Backup Utility program will start. Select **Backup Wizard (Advanced).**

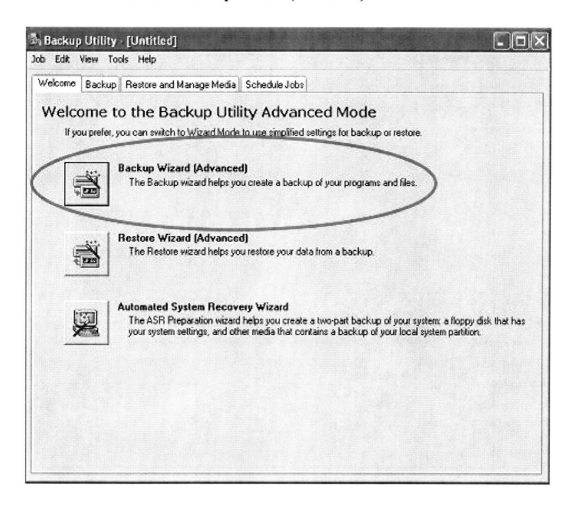

2. The "Welcome to the Backup Wizard" screen will be displayed, click **Next**, then choose **Back up selected files, drives, or network data.** Click **Next.**

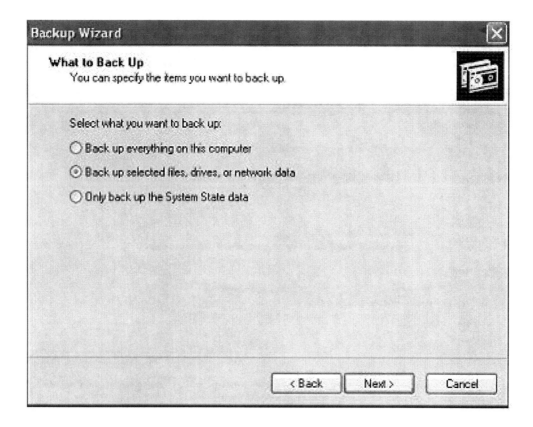

3. In the left-hand pane click on the plus sign next to My Compute to expand the list. Next expand the list under Local Disk (C:). From the left-hand pane, select the folders to backup with a check mark or individual files in the right-hand pane. Once all folders and files are selected, click **Next.**

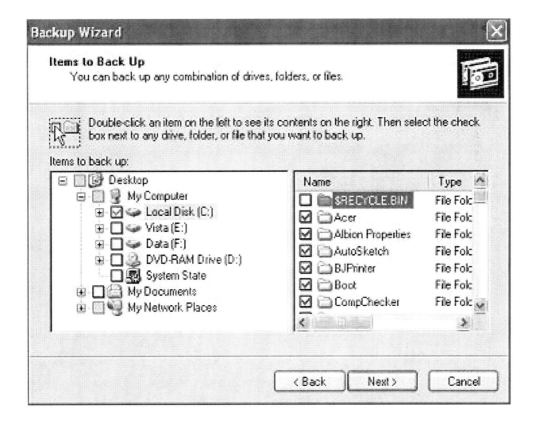

4. On the next screen, leave **Select the backup type** to **File.** Choose the backup destination by clicking on the **Browse...** button. Navigate to My Computer and select the drive of the device. In this example, the backup will be to a Flash Drive labeled **MINI TD (H:)**. (Note: The drive letter may be different for your computer.) Click **Save** to select that location. Give the backup a name in the third box. (The default name is Computer Name_Backup.) Click **Next** to continue.

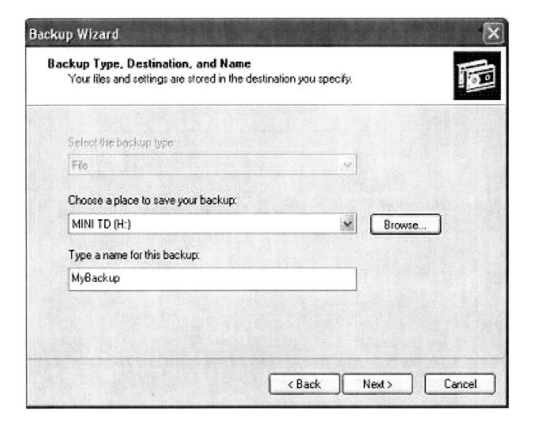

5. A confirmation dialog box will appear. Click **Finish.** The backup will proceed. When it is done, it will give you an option to view the backup report. Click **Close** when done.

Working and Living Independently on the Road   61

Restoring Files with the Windows Backup Software

1. Open the backup program. Click on the **Restore Wizard (Advanced)** button. The "Welcome to the Restore Wizard" screen will be displayed, click **Next** to continue.

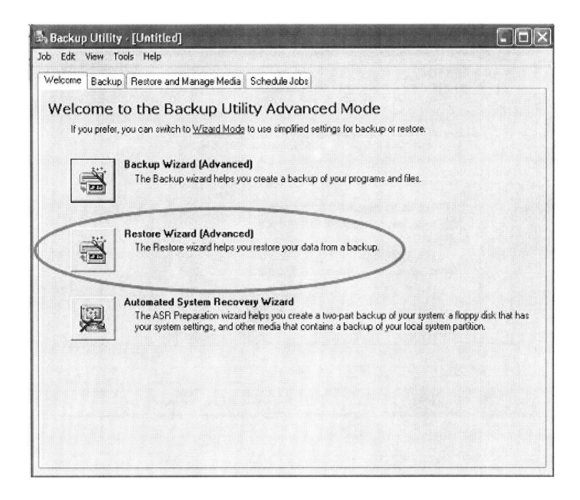

2. The "What To Restore" dialog box will appear. Under the **Items to restore** list in the left-hand pane, expand the list under **File.** Choose the backup (which will have a date next to it) by expanding the list and select the folders and files which you want to restore. Everything with a check mark will be restored. Click **Next** to continue.

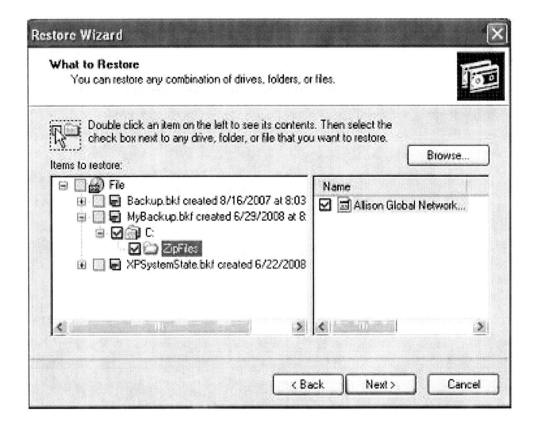

3. On the **Completing the Restore Wizard** dialog box, you may click on the **Advanced** button to choose to restore to the original location (default) or to an alternate location. Click **Finish.** The backup will proceed. When it is done, it will give you an option to view the restore report. Click **Close** when done.

Click on the **Advanced** button to change the restore options. For most file restores, you'll want to replace existing files.

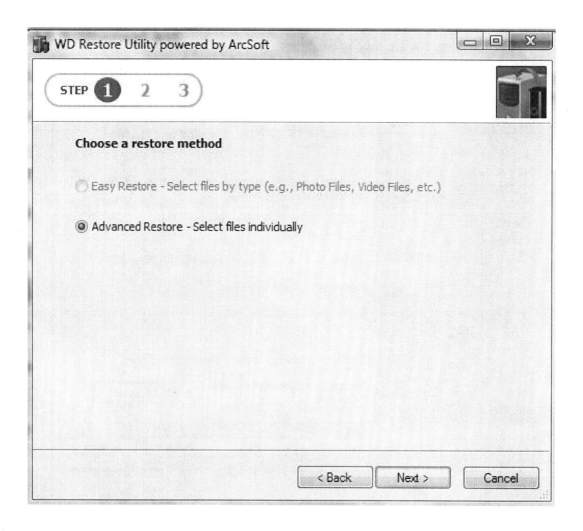

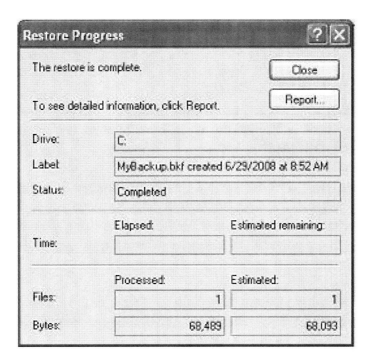

Chapter 3 – The Electrical System

The electrical system in an RV is one of its most important components. All the pleasures, comforts and conveniences of life would not exist without it. The modern mobile office functions on a whole host of appliances that require power.

RV electrical systems aren't complicated, but a thorough understanding of the components and functions is very helpful. The first section will describe electrical basics and equipment that will be handy to have in your tool chest for working on these systems. There will also be some tips on using shore power and what to be aware of in its use. At the end of the chapter, there is a case-in-point example for sizing the electrical storage capacity of the system.

The discussions below will be broken down into AC (Alternating Current) and DC (Direct Current). Some devices in your RV will run on one or the other or both.

THE 120-VOLT AC SYSTEM

AC (Alternating Current)

Let's start with definitions, units of measure and abbreviations that will be used in the text.

The term **Power** will be used frequently and refers to the amount of energy transferred per unit of time. For example, say your RV has a 600-watt microwave oven. The power rating is 600 watts with watts being the unit of power measurement. It will draw (or use) so many **Amperes** of current and will operate at a **Voltage** of 120 volts. There is one simple equation that describes the relationship of amperage, voltage and power:

$$P = V \times A$$

Where:
P is power measured in watts
V is the voltage measured in volts
A is current measured in amperes

In sizing the electrical system, it will be important to know what the rating is for each electrical device. This rating can be found on the U/L tag of the component (see Figure 3.1) and will show its voltage and current information. Knowing the voltage and current will give you the amount of power it uses. (During the planning stage, it may be worthwhile to make a rating list for each electrical component. This will be required in sizing the Solar Power system in chapter 4. If the device does not have a name tag, this information may be obtained off the manufacturer's web site.) This is all the math required – addition, subtraction, multiplication and division.

Figure 3.1 Typical U/L tag for a microwave oven. If the rated voltage is 120 volts AC, and if the oven has a current draw of 8.3 amperes, then power consumption is 1000 watts (1 kW).

In the discussion of the electrical system, you'll need to know some wiring basics and have an understanding of how electrical circuits are designed. When your RV was built, the engineer planned out where the circuits would go and how much load each would carry. An AC circuit-breaker panel was installed to protect each circuit from being over-loaded and the gauge of the wires used were selected to handle the load.

When you add electrical devices to the existing circuits, you need to determine if the circuit has the capacity to handle the additional load. If you add circuits, you'll need to plan for what they will service. This is accomplished by adding up the amperage draw for each device plugged into the circuit that will be operated at the same time.

Example: Circuit 1 has a computer, monitor and printer plugged in. All will be used at the same time. The total amperage is: computer (4 amps), monitor (2 amps), and printer (6 amps), for a total of 12 amps. The circuit must be capable of handling 12 amps or more.

An RV electrical system is designed similarly to household electrical systems. In the example, circuit 1 would be adequate using 14 gauge copper wire and 15 amp protection at the circuit breaker. **IMPORTANT:** The amperage for the circuit breaker or fuse should not exceed the capacity of the wiring in the branch circuit (see Fig3.2).

AWG Copper Wire Gauge *	Amerage Capacity
18	6
16	8
14	15
12	20
10	30
8	55
6	75
4	95
2	130
1	150
0	170

Figure 3.2 Electrical branch circuit design guidelines (For general reference only).

AC Electrical Accessories

Adapters and Extension Cords

Some extra accessories will come in handy when using the RV's 120-volt AC system. There are different types of outlets found in campgrounds and resorts across the country, so having a cache of adapters will prevent frustration when you're trying to hook up to shore power.

Most RVs are equipped with a three-prong, 30-amp plug on the external 120-volt AC power cable, but not all campgrounds have outlets to accommodate this plug. The site may only have 15-amp service (it might be labeled 20-amp) which uses a different plug configuration. For this situation, a 15-amp male to a 30-amp female adapter is required (see Figure 3.3).

Figure 3.3 Different types of 15-amp male to 30-amp female adapters.

Some larger motor homes have 4-prong, 50-amp plugs. Fifty-amp service is 240-volt AC and consists of two separate 120-volt circuits. Adapters used with these plugs combine two 120-volt circuits into one 120-volt circuit. (**Caution:** Do not overload the single circuit when using this type of adapter.) Figure 3.4 shows the three most common campground outlet configurations.

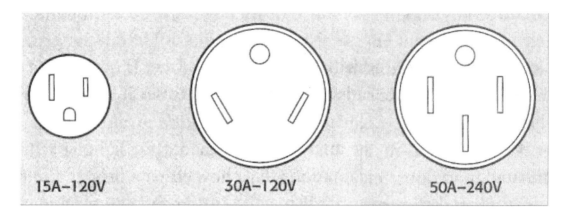

Figure 3.4 The three most common receptacles found at campgrounds.

You may want to include a heavy-duty, outdoor-rated extension cord in a 25 or 50-foot length. Look for the designation of 10/3 which indicates three, 10-gauge wires. These cords are available with 30-amp or 50-amp connectors.

Circuit Analyzers

A circuit analyzer is a valuable tool for checking your RV wiring and campground electrical outlets (see Figure 3.5).

Figure 3.5 Electrical circuit analyzer..

When plugged in, the analyzer's three indicator lights are illuminated in various combinations that show either a proper circuit, a problem with the circuit, or no power at all. An explanation of the light combination is on the housing of the analyzer.

If you are going to need shore power, plug in your RV cable into the 120-volt AC outlet at your site, then check the electric service with the circuit analyzer by plugging it into a 120-volt receptacle in the RV. Do this before plugging in your equipment so as not to cause damage if the electrical service is faulty. If there is a problem, report it to the campsite host and move on to another site.

AC Voltmeters

It's useful to know and monitor how much voltage is received from a campground electrical hookup, generator or inverter. Digital voltmeters are very accurate and easily obtained from the local hardware store. They can give you readings from both an AC and a DC power source.

It's not unheard of to find campsites where the voltage is less that adequate. The campground may be inadequately wired, or the loads being drawn from other campers may be straining the electrical system. Low voltage can do harm to some electrical equipment, causing it to overheat or malfunction. A good rule of thumb is to not operate equipment when below 100 volts.

Surge Protectors

Electronic equipment is very sensitive to voltage surges (spikes). Surges can originate from the local power plant and also by electrical storms when lightning strikes power lines. To guard against spikes, use a surge protector (see Figure 3.6).

While surge protectors are not lightning arresters, they offer some protection from lightning. It is a good idea to buy a surge protector that also has coaxial connections that can be used to protect your television equipment, and telephone jacks that can protect your computer modem.

Figure 3.6 Surge protector with widely spaced receptacles for plugging in transformer boxes, coaxial connections for TV/Satellite and RJ11 telephone jacks to protect computer modems and phone connections. (Courtesy Monster® Cable Products http://www.monstercable.com)

THE 12-VOLT DC SYSTEM

DC (Direct Current)

The same formula for calculating power applies for DC: **P = V x A.**

The 12-volt DC system is the primary electrical system since most items in an RV run off DC power (Interior/exterior lights, water pump, furnace, fuse panel or circuit-breaker panel). In addition, some RVs are equipped with refrigerators that can be operated on a 12-volt current, and may have certain other items such as a stereo that runs off the 12-volt system.

With the installation of 12-volt outlets, an AC/DC TV and other 12-volt appliances and equipment such as VCRs, DVD players, fans and a laptop (equipped with a converter), can be operated from the RV's batteries.

The batteries are a very important component of the DC electrical system. See the section called <u>The Battery System</u> later on in the book for a full discussion on the battery system and components. The typical RV will have one or two "house" batteries (deep-cycle batteries) and one or two automotive type batteries (SLI – Starting, Lights, Ignition) for automotive functions. I will limit our discussion to the house batteries.

Batteries are the storage medium for powering all the DC powered equipment during the time that you are independent of shore power. The size, type and quantity of the batteries is dependent on how long you wish to operate any of this equipment without hookup to shore power or the running of a generator. Many RV owners assume that they are always going to hook into shore power while camping or will only power the DC equipment for short periods of time. I am going to go into detail about designing a system that will allow you to be more independent of shore power (and/or a generator) so that your equipment can function for days at a time on the batteries alone.

<u>Power Load Requirements</u>

The starting point for sizing your DC system is to determine the power requirements of both the DC and AC appliances and equipment in your RV (DC power can be converted to AC power with an inverter, which will be discussed in detail later).

Earlier I discussed how to find the power requirements off the name plate on the device. You will need to gather this information to run a calculation of your power needs. I have provided a "typical" hourly amperage draw for a number of common items found in an RV (see Figure 3.7). I suggest you use the power tag information from the item when possible to give you a more accurate calculation, or purchase an ammeter to monitor the power usage (see Figure 3.8).

Item	Amps
TV, Color, 9-inch (DC)	4.0
TV, B&W, 9-inch (DC)	2.0
Light, incandesent (DC)	1.2
Light, single fluorescent (DC)	0.7
Furnace, 31,000 BTU	8.2
Microwave, 600 watt (AC)	5.0
Laptop computer (DC)	3.2
Printer (AC)	1.0
Scanner (AC)	1.1
19" LCD Monitor (AC)	2.5
VCR (AC)	0.4
CD Player (AC)	0.3
Stereo (AC)	0.3
Satellite Dish AC)	0.3
CB Radio (DC)	0.1
Water Pump (DC)	4.6
Fan (DC)	1.0

Figure 3.7 Typical hourly amperage draw of AC and DC RV equipment.

Figure 3.8 The kill-a-watt meter for measuring the power usage of individual appliances and equipment. (Courtesy P3 International http://www.p3international.com)

...o think about how many hours a day each of the electrical devices ...imple, you might use your laptop for 4 hours/day, the TV for 3 ...a table (see Figure 3.9) so that you can tally up the estimated total ...er day. This information will then be used to compute how many ...batteries between charging times. Note that laptops have their ...... ue used without being plugged in, but the recharging will consume more Amp-Hours than running the laptop off the house batteries.

Item	Amps	Time Used	Amp-Hours Consumed
Laptop	3.2	4.0	12.8
Monitor	2.5	4.0	10.0
Printer	1.0	0.5	0.5
TV	3.0	3.0	9.0
Fan	1.0	2.0	2.0
3 Lights	3.6	3.0	10.8
Water pump	4.6	0.5	2.3
Misc.	2.0	2.0	4.0
		Total:	51.4

Figure 3.9 Daily power requirements.

Battery capacity is measured in Amp-Hours, therefore the table in Figure 3.9 has a column for number of Amp-Hours consumed. (AH = A x Time) You will now have a rough estimate of electrical usage so that you can settle on how much battery capacity to purchase.

The Battery System

Motor homes have two types of batteries: the deep-cycle type for house use, and another known as the SLI (Starting, Lights, Ignition) type for automotive functions.

SLI batteries are usually the no maintenance type that need to be kept fully charged. See sidebar below for a more detailed explanation of battery types. This is accomplished by the engine alternator like any normal car battery is charged. SLI batteries should not be used as house batteries since they are not designed to have a large portion of their capacity drained before they are recharged.

Battery Types and Terminology

Batteries are divided in two ways, by application and construction. The major applications are automotive, deep-cycle and marine. Deep-cycle batteries uses include solar electricity, backup power, and RV and boat "house" batteries. The major construction types are flooded (wet), gelled, and AGM (Absorbed Glass Mat).

Battery Application

Automotive or Starting (sometimes called SLI, for Starting, Lighting, Ignition) batteries are commonly used to start and run engines. Engine starters need a very large starting current for a very short time. Starting batteries are designed with a large number of thin lead plates for maximum surface area. The plates are composed of a lead sponge which gives a very large surface area, but if deep-cycled, the sponge will quickly be consumed and fall to the bottom of the cells. Automotive batteries will generally fail after a few hundred deep cycles, but will last for thousands of cycles in normal starting use where the battery discharges only a few percent of its capacity.

Deep-cycle batteries are designed to discharge as much as 80% of their capacity for thousands of cycles. They are made with thick solid lead plates verses the lead sponge design of automotive batteries. They have less surface area which means that the battery has less instant power than the starter battery, but provides a much longer battery life.

Marine batteries are typically a "hybrid", and fall between the automotive and deep-cycle batteries. In the hybrid, the plates may be composed of a lead sponge, but it is coarser and heavier than that used in starting batteries. "Hybrid" types should not discharge more than 50% of their capacity.

Battery Types and Terminology (Cont.)

Battery Construction

Flooded or wet-cell batteries (the ones with the removable caps) have lead plates and are filled with an electrolyte (usually acidic, but sometimes alkaline) solution, which needs to be periodically replenished with distilled water. When batteries are termed "sealed," it means that they are made with vents that (usually) cannot be removed. Maintenance Free batteries are also sealed, don't need electrolyte refills, but are not usually leak proof. Sealed batteries are not totally sealed, because they must allow gas to vent during charging.

Gelled batteries, or "Gel Cells," are sealed and contain acid that has been "gelled" by the addition of silica gel, turning the acid into a solid mass. The advantage of these batteries is that it is impossible to spill acid, even if they are broken. However, there are several disadvantages. One is that they must be charged at a slower rate to prevent excess gas from damaging the cells. They cannot be fast charged on a conventional automotive charger or they may be permanently damaged. This is not usually a problem with solar electric systems, but if an auxiliary generator or inverter bulk charger is used, current must be limited to the manufacturer's specifications. Most better inverters commonly used in solar electric systems can be set to limit charging current to the batteries. Another disadvantage of gel cells is that they must be charged at a lower voltage than flooded or AGM batteries. If overcharged, voids can develop in the gel, causing a loss in battery capacity.

Sealed AGM (Absorbed Glass Mat) batteries use a very fine fiber, Boron-Silicate glass Mat between the plates. The mat is about 95% saturated with an acidic electrolyte solution, rather than fully soaked. (This type is also known as a "starved electrolyte" or "dry" battery.) This means that the battery will not leak acid even if broken, since all the electrolyte solution is contained in the glass mats. The charging voltages for AGM batteries are the same as for any standard battery, so there is no need for any special adjustments or problems with incompatible chargers or charge controllers. AGM batteries have the main advantages of no maintenance, being completely sealed against hydrogen leakage, non-spilling even if broken, and the ability to survive most freezes. The main disadvantage is that they cost 2 to 3 times more than a flooded, wet-cell battery.

Source: Northern Arizona Wind & Sun. (http://www.windsun.com)

House batteries, the deep-cycle type, are the only type suitable for RV house use. Deep-cycle batteries are also referred to as marine-use batteries. These batteries can be the wet-cell type that needs the periodic addition of water, or the maintenance-free type.

Deep-cycle batteries come in a variety of sizes. Many have "group" sizes which are based upon the physical size and the terminal placement, both of which are not a measure of battery capacity. Typical batteries in motor homes are classified as group 24, 27, 29 or 31. See Figure 3.10 for approximate battery capacity. When adding additional batteries to your RV, you will need to be somewhat creative in finding a place to put them. Most motor homes usually have only one or two battery trays designed specifically to hold the battery from moving and to provide ventilation. (It is **very important** to note that wet-cell batteries MUST be ventilated because they may emit hydrogen gasses while being charged, which are explosive.) Batteries can also develop leaks if damaged. Fortunately, batteries are now available that are completely sealed, don't leak gasses and will not leak even if broken. See sidebar above regarding battery types. These batteries are more expensive than the wet-cell type, but may be required depending on where you will place the battery.

Group size	Capacity (Amp-Hous)	Voltage
24	70-85	12
27	85-105	12
29	105-117	12
31	95-125	12

Figure 3.10 Common battery group sizes and approximate capacities.

Battery Ratings

All deep cycle batteries are rated in amp-hours (AH). An amp-hour is the number of hours a battery can be used before it needs to be charged again and is defined as one amp (A) for one hour, or 10 amps for 1/10 of an hour, etc. The formula is AH = amps x hours. For example, if you have an electrical device that draws 15 amps, and you use it for 30 minutes, then the AH used would be 15 A x 0.50 hr, or 7.50 AH. The accepted AH rating time period for most all deep-cycle batteries is the "20-hour rate." This means that the battery is discharged from a 100% state-of-charge (full) down to 10.5 volts (0% state-of-charge, or empty) over a 20-hour period, while the total actual amp-hours it supplies is measured during the test with an amp-hour meter. Sometimes you will see different ratings such as the "100-hour rate" for comparison purposes. This means that the battery is discharged from a 100% state-of-charge (full) down to 10.5 volts (0% state-of-charge, or empty) over a 100-hour period, while the total actual amp-hours it supplies is measured during the test with an amp-hour meter. Therefore, make sure to compare batteries at the same discharge rate to see if the battery will meet your needs..

Battery amp-hour ratings are specified at a particular rate because of the Peukert Effect. The Peukert value says that the higher the internal resistance, the higher the losses while charging and discharging, especially at higher currents. This is directly related to the internal resistance of the battery. This means that the faster a battery is discharged, the lower it's AH capacity. On the other hand, if it is drained more slowly, the AH capacity is higher.

It should be pointed out that as a battery is discharged, the voltage decreases. When the voltage drops below 11.8 volts, a good deal of electrical equipment will cease to operate. Many batteries reach this voltage when they have been discharged to below 50 percent of their capacity. Therefore, it is recommended to never discharge your house batteries below this level.

When you add batteries to the existing bank of batteries in the RV, it recommended that the batteries be of the same size, age and from the same manufacturer. (Connect the batteries with #2 minimum gauge copper wire. The batteries should be connected in parallel (see Figure 3.11) for a 12-volt system. Each additional battery will add to the total Amp-Hour capacity of the system. (If you started with one battery rated at 100 AH and added a second 100 AH battery, you would have 200 AH of capacity.)

There are two additional factors that affect battery capacity – temperature and battery efficiency. A battery's capacity is typically rated at 77 °F (25 °C). As the operating temperature of the battery decreases, it's efficiency decreases (see Figure 3.12) and this must be taken into account to properly size the battery system. Also, battery efficiency deteriorates slightly over time.

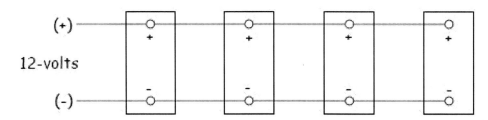

Four 12-volt batteries wired in parallel. The overall voltage remains the same but the capacity is four times that of one battery. Therefore, if each battery has a rating of 100 AH, the total capcacity is 400 AH.

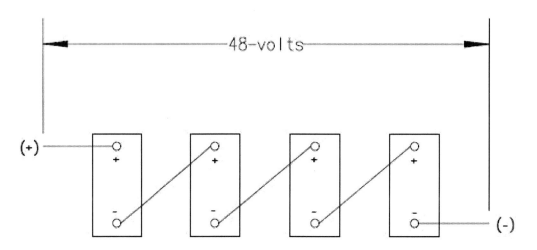

Four 12-volt batteries wired in series. Batteries connected in series have the positive terminal of one battery connected to the negative terminal of another battery. In this example, the output voltage would be 48-volts. If each battery has a rating of 100 AH, the total capcacity is still 100 AH.

Figure 3.11 Battery connection arrangements. Shown are how connecting batteries in parallel or in series is accomplished. Connecting 12-volt batteries in parallel provides 12-volt power, while connecting batteries in series allows you step up the voltage produced based on the number of batteries in the series (such as 24-volt, 48-volt).

Ambient Temperature	Factor
80F	1.00
70F	0.96
60F	0.90
50F	0.84
40F	0.77
30F	0.71
20F	0.63

Figure 3.12 Effect of temperature on battery efficiency.

Battery Charging

An important part of the overall battery system is the method in which the batteries will be recharged. Most RVs are designed to have the house batteries recharged from the engine alternator while the engine is running. And if the RV is equipped with an inverter (see inverter section below) that has a built-in battery charger, the batteries will be charged automatically while the RV is connected to shore power. If there is not a built-in charger in the Inverter, a separate battery charger could be purchased and used while shore power is available. A third method of battery charging is via solar electrical panels, which will be covered in detail in chapter 4.

12-Volt DC Outlets

Cigarette-lighter plugs and sockets are what the RV industry uses as 12-volt electrical connections (see Figure 3.13). These also come with safety plug covers so that you can't accidentally insert your finger and get shocked. See Figure 3.14. If you find the appearance of the cigarette-lighter plug style unsightly, you can purchase the smaller version of the standard 120-volt AC outlet that use two-prong plugs. See Figure 3.15.

Figure 3.13 Example of a 12-volt electrical receptacle. (Courtesy Camping World http://www.campingworld.com)

Figure 3.14 An example of a 12-volt electrical receptacle with safety cap. (Courtesy Camping World http://www.campingworld.com)

Figure 3.15 Example of a 12-volt electrical receptacle. (Courtesy Camping World http://www.campingworld.com)

Either way you go, you will need to have 12-volt connections available for your 12-volt appliances and equipment. Since many motor homes don't come equipped with any 12-volt connections (or there are not enough, or they are located in the wrong place), you may find the need to add these to your RV.

DC-to-AC Inverters

Not all of your equipment and appliances are going to run on DC power. In fact, most will run only on AC power unless you specifically purchase them to use 12-volts DC (see Figure 3.16). Therefore, an inverter is required to convert the 12-volt DC battery power into 120-volt AC power (see Figure 3.17). Most modern mobile homes come standard with an inverter, but of course you can also find them at numerous shops on-line.

Figure 3.16 An example of a 12-volt toaster oven. (Courtesy Sports Imports LTD http://www.sportsimportsltd.com)

Figure 3.17 Samlex 600 watt pure sine wave 12-volt inverter. (Courtesy Samlex America http://www.samlexamerica.com)

Inverters, with well thought-out usage, are a valuable accessory. Unlike a generator that produces AC power, they run silently. Inverters are best used for short-term, 120-volt AC convenience loads, such as running appliances that aren't normally run for long periods. Another good usage of the inverter is powering laptop computers and monitors, since they draw small amounts of current. When shopping for an inverter you will find that many choices confront you. With prices ranging from less that $40 to well into the thousands, it can be difficult to know what features are important and how to choose a unit appropriate to your needs. While wattage, features, and connections are obvious considerations, there are other factors such as the inverter's wave form output. See sidebar below which goes into more detail. In general, look for a pure sine wave inverter to run sensitive equipment like computers, monitors, flat panel TVs, and printers. Choose what is known as a modified-sine wave (or square wave) inverter to run motors such as those used in electric drills, shavers, etc.

Keep in mind that inverters use a good deal of power to make the conversion. A rule of thumb to remember is that roughly 10 amps of battery current will be drawn for every amp of AC power produced. As an example, we'll use a 240-watt piece of equipment, and we'll run it for 3 hours off the inverter. To figure the number of amp-hours consumed during that time, take 240 watts divided by 12 volts which gives us 20 amps used in one hour. Running it for 3 hours will use 20 times 3, or 60 amp-hours from the battery system. In addition, to more accurately calculate the usage, the inverter's efficiency must be taken into account. Typical inverters have an efficiency of 90%, so the calculation above needs to be multiplied by 1.1 (1.0/0.9) which gives us a total amp draw of 66 amp-hours (or 22 amps per hour).

To properly size the inverter, simply add up the power (wattage) used by all the equipment that will be run simultaneously (remember conversion formula: Watts ÷ 12 = Amps .... Amps x 12 = Watts) and add 20% as your minimum power requirement. Keep in mind that tools with induction motors (e.g. saber saw, portable grinder, air compressor) may require 3 to 7 times the listed wattage when starting. The start-up load of the appliance or tool determines whether the inverter has the capability to power it. Be sure to check the specific wattage requirements and operating instructions for the appliances and equipment to be used.

Once you have calculated your inverter load requirements, check the specification sheet of the inverter that was installed in the RV to be sure it is adequate for your needs. If your RV did not come equipped with one, use the guidelines above to properly size one.

Inverter wave form output – modified sine-wave vs. pure sine-wave

When choosing an inverter, one of the considerations is the wave form it puts out. Besides the power rating on an inverter, it will be classified as either **Modified Sine-Wave** or as a **Pure Sine-Wave**. The distinction between the two is in the quality of the output voltage.

Pure sine-wave output is similar to the clean power provided by the utility companies which have very low harmonic distortion and a pure sinusoidal wave form factor.

[INSERT Graphic (File: InverterWaveForms.jpg)]

This type of output is mandatory for most electronic equipment (computers, printers, monitors, flat panel TV, CD/DVD players, game consoles, audio amplifiers, etc.), solid state power tools, battery chargers, and medical equipment like oxygen concentrators. Pure sine-wave output also helps inductive loads like microwave ovens and motors run faster, quieter and cooler. The disadvantage of this type of inverter is that it is considerably more expensive.

The modified sine-wave, as shown above, produces a clipped wave form but still supplies the same output voltage as the pure sine-wave design. For less sensitive electrical needs, it is the more economical choice.

System Monitoring

If you want to be independent of shore power hookups and use your batteries as the primary power source, having an accurate method of checking the state-of-charge and checking out the electrical system in general is very useful. This can be done with simple multimeters, panel meters and monitor panels.

At a minimum, a voltmeter should be purchased for monitoring the batteries (see Figure 3.18). It can indicate the battery state-of-charge, problems with the charging system and problems that may be developing in the battery. The typical RV has a built-in monitor panel which usually includes a battery-condition meter which has a simple meter with no markings or a series of lights. Either type is quite useless and should not be relied upon.

Figure 3.18 Panel type voltmeter for monitoring battery conditions. (Courtesy Radio Shack http://www.radioshack.com)

In addition to a voltmeter, it would be a wise investment to purchase a good quality digital multimeter (see Figure 3.19) that will function as an even more accurate voltage meter and provide other functions for troubleshooting your AC and DC electrical systems and equipment. Make sure the readout on the digital meter measures to two decimal places. The reason is that the difference between a fully-charged battery and a completely discharged battery is about 0.9 volts (see Figure 3.20). A two-decimal place readout provides 90 measuring increments whereas a one decimal place readout only provides 9 increments. An instrument with more increments indicates faster up or down changes in voltage and therefore gives the most information and provides a more accurate reading of voltage while charging. Please note that to get the most accurate battery voltage reading, read the voltage after the battery has rested for at least fifteen minutes with no discharge or charge applied.

Figure 3.19 A multimeter with digital readout to provide accurate measurements. (Courtesy Radio Shack http://www.radioshack.com)

% of Charge	Battery Voltage
100	12.7
95	12.64
90	12.58
85	12.52
80	12.46
75	12.4
70	12.36
65	12.32
60	12.28
55	12.24
50	12.2
40	12.12
30	12.04
20	11.98
10	11.94

Figure 3.20 State-of-Charge of a 12-volt battery system by voltage.

To really get the "big picture" on the status of your battery system, a mid-priced panel meter can keep track of the energy your system has available, as well as the energy consumed. This can ensure adequate reserve power capacity, as well as ensure the longevity of the battery bank (see Figure 3.21).

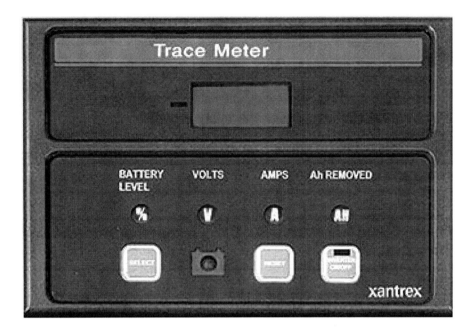

Figure 3.21 Panel-style digital meter to accurately keep track of your battery bank status. (Courtesy Xantrex - http://www.xantrex.com)

Panel meters provide easy-to-understand information such as the produced and used DC current. They can keep track of the energy your system has available and how much energy has been used. These meters can be thought of as your energy supply "fuel gauge," used in the same way you use your RV's gasoline fuel gauge. These meters are sophisticated energy storage computers that record battery system characteristics and data. The example meter shown in Figure 3.21 is capable of measuring the battery voltage, net battery current draw, the battery's charge level, the accumulated amp-hours used and the number of days since the battery bank was fully charged. Having this information at hand is very practical in determining battery status, estimating battery life, and proper battery management. For serious RVs, it's the way to go.

Battery Maintenance

Safety Warnings:

Flooded (wet-cell) batteries contain a mixture of acid and water. Avoid skin contact by wearing protective clothing, gloves and safety glasses.

> Never add acid to a battery.

> Do not over tighten battery terminals. Doing so can result in post breakage, post meltdown and fire.

> Good battery maintenance will help prolong the life of your batteries and should be part of your overall RV maintenance plan. Your batteries should be inspected on a regular basis. Things to look for include:

- Examine the outside appearance of the battery. Look for cracks, budges or other damage (repair or replace damaged batteries).
- Make sure the top of the battery, the posts, and the connections are clean, and free of dirt, fluids, or corrosion.
- Check the battery cables and their connections. Broken or frayed cables can be extremely hazardous. Replace any cable that looks suspicious.
- Tighten all wiring connections. Make sure there is good contact with the terminal posts.

For flooded-type batteries (the ones with the removable caps), check the water level at least once a month until you get an idea of how thirsty your batteries are. The minimum water level is at the top of the plates. Things to avoid when watering batteries:

- Don't let the plates get exposed to air because it will corrode the plate.
- Don't overfill the battery cells. Add water to 1/8" below the bottom of the fill well.
- Don't use tap water. Use distilled water only.

Since batteries seem to attract dust, dirt, and grime, keep them clean. Here's a cleaning check list:

- First check that all vent caps are tightly in place.
- Clean the top of the battery with a cloth or brush and a solution of baking soda and water (i.e. a teaspoon of baking soda dissolved in a small 3-ounce paper cup filled with water).
- When cleaning, don't allow any cleaning solutions to get inside of battery.
- Rinse with water and dry with a clean cloth.
- Apply a thin coat of petroleum jelly on the terminals to prevent corrosion.

Finally, here are a few helpful hints about battery management:

- Shallow discharges will result in a longer battery life.
- 50% (or less) discharge level is recommended.
- 80% discharge is the maximum safe discharge level.
- Never fully discharge flooded (wet cell) batteries since this will damage the battery.
- Do not leave batteries deeply discharged for any length of time.
- Lead-acid batteries do not develop a memory and need not be fully discharged before recharging.
- Batteries should be charged after each period of use.

A CASE-IN-POINT EXAMPLE

Now that the electrical basics have been covered, I'd like to conclude this chapter with a step-by-step discussion on sizing the battery system and calculating the length of time the system will support your electrical needs without being connected to shore power.

Step one is to determine your power requirements. I've created an example table of items that require power and their estimated daily usage (see Table 3.1). The power figures can be obtained from either the name plate information, an estimated power usage table (see Figure 3.7), or from an ammeter.

Item	Amps	Time Used	Amp-Hours Consumed
Laptop	3.2	4.0	12.8
Monitor	2.5	4.0	10.0
Printer	1.0	0.5	0.5
TV	3.0	2.0	6.0
Fan	1.0	3.0	3.0
2 Lights	2.4	3.0	7.2
Water pump	4.6	0.5	2.3
Furnace	8.2	1.0	8.2
		Total:	50.0

Table 3-1 Example of daily power requirements.

Step two is to take the power requirements above and factor in the depth of allowable battery discharge and the effect of temperature (70 °F in this example). This will yield the corrected total amp-hours required per day.

1. Daily amp-hour required = **50 AH**
2. Allowable battery discharge level = **50%**
3. Divide line 1 by line 2 (50 AH / 0.50) = **100 AH**
4. Correction factor for battery temperature of 70 °F (see Figure 3.12) = **0.96**
5. Divide line 3 by line 4 (100 AH / 0.96) = **104 AH**

Step three is to calculate the number of batteries required to power the load between battery recharging. In this example I have chosen the Concorde PVX-1080T battery which is rated at 108 AH (www.concordebattery.com) and I wish to have two full days between charging. Therefore, based on a battery rating of 108 AH and 2 full days of usage (208 AH), it will require 208 AH / 108 AH = 1.93 batteries. Rounding the result upwards indicates that 2 batteries are needed.

The example above assumes no battery recharging during two consecutive days, but most likely there will be recharging taking place if you drive the RV during that time or plug into shore power.

For those who wish to be independent of shore power and run on batteries for longer periods of time, solar power may be an option that you might want to explore. This is the topic of the next chapter.

Chapter 4 – Solar Power

One of the best methods of battery charging is with solar panels. A solar electric system for an RV is quiet, light weight, requires no operator attention and requires little or no maintenance.

IS SOLAR RIGHT FOR YOU?

Adding solar power to your motor home is obviously going to cost additional dollars to install. So the question is whether it is worth the additional expense and time. What are the benefits?

- It's quiet. No need to be bothered by the noise of a generator or engine running when the batteries need to be recharged.
- Reduces dependence on finding shore power and gives you the freedom to park anywhere and still have power available.
- Reduces the smell and pollution of running fossil fuel engines to generate power.
- It is the ideal way to recharge your battery bank since it applies a trickle charge and keeps the batteries in the fully charged state a higher percentage of the time. This prolongs the life of your batteries.
- It saves you the cost of paying more for a camping site that has electric power available.
- It saves energy!

Something else to consider is that the return on investment for solar power is favorable. The investment can generate a quick payback as well as long term savings. Consider this example: You purchase the necessary equipment to provide solar electricity (solar panels, solar regulator, cables, etc.) for $2,500. Knowing that state parks and private RV parks and campgrounds charge an average of $5 extra for electrical hookups, you can factor in the savings obtained by staying at non-electrical sites or paying the per kilowatt charge.

What are the savings? Obviously the more nights a year you camp out, the faster the return on your investment. Here are a couple scenarios:

Full-time RV user camping 365 days a year: The daily savings is $5 multiplied by 365 days for an annual savings of $1,825. The ROI (Return On Investment) is $2,500 divided by $1,825 per year, or 1.37 years (or approximately 1 year, 5 months).

Half-time RV user camping 180 days a year: The daily savings is $5 multiplied by 180 days for an annual savings of $900. The ROI (Return On Investment) is $2,500 divided by $900 per year, or 2.77 years (or approximately 2 years, 9 months).

As you can see, the good returns are realized by the RV enthusiast or serious business traveler.

So if solar power interests you, please read on.

SYSTEM COMPONENTS

Solar power systems can be as simple as connecting a small solar panel directly to a battery and producing a few amp-hours per day, and can cost less than $100. In this book, I'm going to focus on the more sophisticated systems that consist of multiple solar panels and solar charge controllers that produce enough power to run a microwave oven.
The following is a discussion of the four primary components for producing electricity from solar power.

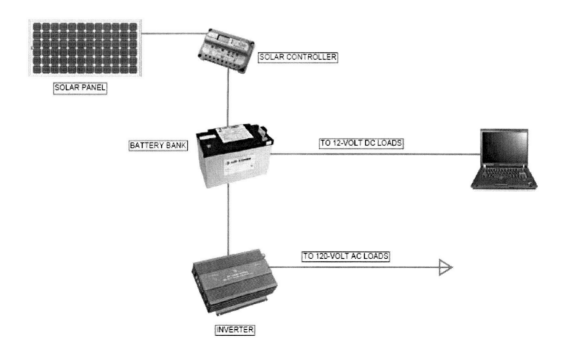

Solar Panels

Solar panels, also referred to as photovoltaic cells (PV cells), are the heart of the system. The solar panel's photoelectric cells collect light from the sun and convert it to electrical energy. Panels come in a wide variety of sizes and shapes so you can tailor your system to meet the power requirements you need and fit them onto the space available on the RV rooftop. Solar power for mobile homes has become so popular that many stores sell pre-packaged kits that take the guesswork out of the selection process. But since every installation is somewhat different, it is good to know what goes into a complete solar power kit.

Solar panels

A solar panel is a device that collects and converts solar energy into electricity or heat. Solar photovoltaic panels are made so that the sun's energy excites the atoms in a silicon layer between two protector panels. Electrons from these excited atoms form an electric current that can go directly to a device to supply power, like a calculator, or to a storage device, like a battery.

The basic element of solar panels is pure silicon. When stripped of impurities, silicon makes an ideal neutral platform for the transmission of electrons. In silicon's natural state, it carries four electrons, but it has room for eight. Therefore, silicon has room for four more electrons. If a silicon atom comes in contact with another silicon atom, each receives the other atom's four electrons. Each atom achieves the "ideal" eight valence electrons; this creates a strong bond, but there is no positive or negative charge. This material is used on the plates of solar panels. Combining silicon with other elements that have a positive or negative charge can also create solar panels. [1]

For example, phosphorus has five electrons to offer to other atoms. If silicon and phosphorus are combined chemically, the results are a stable eight electrons with an additional free electron. The silicon does not need the free electron, but it cannot leave because it is bonded to the other phosphorous atom. Therefore, a silicon and phosphorus plate is considered to be negatively charged.[1]

A positively charged solar panel can also be created which allows electricity to flow. Combining silicon with an element such as boron, which only has three electrons to offer, creates a positive charge. A silicon and boron plate still has one spot available for another electron. Therefore, the plate has a positive charge. The two plates are sandwiched together to make solar panels, with conductive wires running between them.[1]

Photons bombard the silicon/phosphorus atoms when the negative plates of solar cells are pointed at the sun. Eventually, the 9th electron is knocked off the outer ring. Since the positive silicon/boron plate draws it into the open spot on

its own outer band, this electron doesn't remain free for long. As the sun's photons break off more electrons, electricity is then generated. When all of the conductive wires draw the free electrons away from the plates, there is enough electricity to power low amperage motors or other
electronics, even though the electricity generated by one solar cell is not very impressive by itself. When electrons are not used or lost to the air they are returned to the negative plate and the entire process begins again.[1]

[1] Pollick, Michael. How Do Solar Panels Work?. Wisegeek. Conjecture Corporation.. Retrieved on 2008-05-07.

Solar panels are sold by watt rating. For example, a 65-watt rated panel will produce approximately 350 watts in 5 hours of sunshine. If you check the specification sheet for the panel, it will denote the electrical performance, and the maximum power voltage and current at a given test condition. For example, a solar panel may have an irradiance of $1000W/m^2$ at a panel temperature of $25°C$. If the panel produces 4 amps, you would have 4 amps available to run your equipment or to recharge a battery. With two panels of this size wired in parallel, you would have 8 amps available.

One of the nice features of solar panels is that they're easy to install and hook together. There is a wide selection of mounting kits that provide the means to mount the panels to the roof. Simple angle brackets are sold to mount the collector flush with the roof or there are mounting rails that allow you to position the panel at an angle to best capture the sun's rays or lay flat while traveling. These are typically fastened to the roof with 8 mounting screws and some caulk to weather seal around the screw holes. Each module has a sealed electrical junction box which you connect together with 10 or 12 gauge inter-module cables. If you need to expand the system, purchase more panels and wire them in parallel (for a 12-volt DC system) with the other solar collectors to give you more power output.

<u>Solar Panel Regulators</u>

Since the solar panels are being used to recharge the battery bank, a method to control (regulate) the charging voltage is required to properly charge the batteries and prevent overcharging (see Figure 4.1). Directly connecting a solar array to a battery bank with no regulation will compromise the life of the battery.

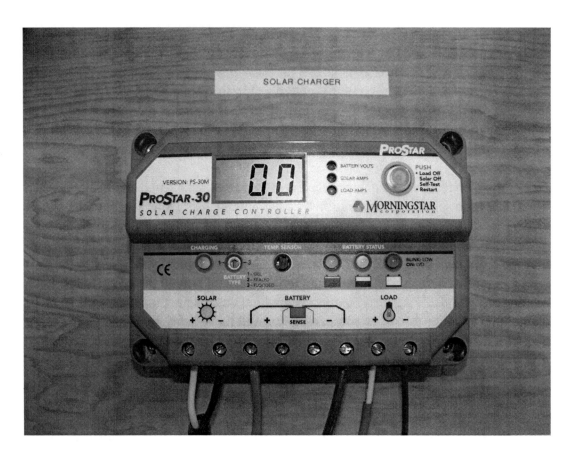

Figure 4.1 Solar charge controller. (Courtesy of Morningstar Corporation http://www.morningstar.com)

Solar charge controllers

The main function of a solar controller or regulator is to fully charge a battery without permitting overcharge, while also preventing the reverse current flow at night.

The circuitry in a controller reads the voltage of the batteries to determine the stateof-charge. Designs and circuits vary, but most controllers read voltage to control the amount of current flowing into the battery as the battery nears full charge.

Features to look for in a good controller are:

- Reverse current leakage protection – this is accomplished by disconnecting the array or using a blocking diode to prevent current loss into the solar modules at night.
- Low-voltage load disconnect (LVD) – this reduces the damage to batteries by avoiding deep discharge.
- System monitoring – this is done with analog or digital meters, and indicator lights and/or warning alarms.
- Overcurrent protection – this is accomplished with fuses and/or circuit breakers
- System control of other components in the system, such as standby generator or auxiliary charging system.
- Pulse Width Modulation (PWM) – an efficient charging method that maintains a battery at its maximum state-of-charge.

Charge controllers are purchased by amperage capacity (the solar array current produced at maximum sunlight) and system voltage required. For example, if one solar module in your 12-volt system produces 4.5 amps and 2 modules are utilized, your system will produce a maximum of 9 amps of current at 12-volts. It is recommended by most controller manufacturers and installers to build in a 25% safety margin when sizing the controller. In this example, that would bring the minimum controller amperage to 11.3 amps. Looking through the selection of products, a 15-amp controller is the closest match. Obviously there is no problem going with a 20- or 30-amp controller other than additional cost. If you think that in the future you may expand the system, additional amperage capacity should be considered.

Once you have the controller size established, look at the available features. For a relatively small increase in cost, you can purchase units that have LED readouts that display such things as battery voltage, amperage being produced and DC load. These features are highly recommended since they give you real-time system performance and help in trouble-shooting the solar electrical system. They will also display the battery state-of-charge along with the in-progress charging mode.

Installation of the solar regulator is straightforward. Connect the battery bank first, using at least 12 GA copper wire (10 GA is preferred to reduce power loss in the transmission between the solar panel and the battery). Next, connect the solar panels to the solar regulator using at least 12 GA copper wire (again, 10 GA is preferred). (**Safety Precaution:** During the connection of the solar panels, it is highly recommended to cover the solar panel or place it out of direct sunlight so that no current is being produced. A more elegant solution is to have a disconnect switch between the panel and controller that can be thrown to the OFF position while installing or working on the system.)

Finally, connect any DC loads that you wish to run off the solar electrical system directly. If you plan on having multiple different DC-powered equipment running off the panels, consider installing a load distribution box and connect the loads from there. This load center can also house the fuses that protect the equipment from an amperage overload. All DC loads should have a fuse installed between the power source and the device being powered. Check the manufacturer's literature for the proper fuse size and type.

Direct Current Fuses

DC (direct current) fuses come in a variety of shapes and sizes. The most common types you should be familiar with as applied to the topics in this book are the glass and ceramic cartridge fuses, the automotive-style blade type plug-in fuses (ATO or ATC), and the industrial Bussmann ANN fuses.

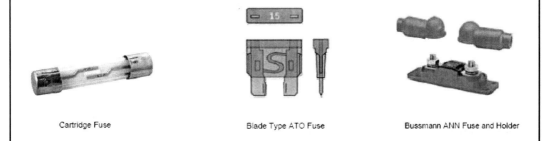

Cartridge Fuse   Blade Type ATO Fuse   Bussmann ANN Fuse and Holder

The cartridge fuse is designed with either a glass tube or a ceramic tube body between the metal end caps with a thin metal filament wire connecting both ends. Glass fuses have the advantage of the fuse element being visible for inspection purposes, but they have a low breaking capacity which generally restricts them to applications of 15 A or less at 250 V. Ceramic fuses have the advantage of a higher breaking capacity, facilitating their use in higher voltage/ampere circuits. The wire is designed to break (or "blow") when the current exceeds the fuse's rating.

Blade type ATO (Automotive Technology Organization) or ATC fuses have a plastic body that houses the fuse element and two prongs that fit into the connection sockets of the fuse holder. And like the cartridge fuse, the wire is designed to break when the current exceeds the fuse's rating. Blade fuses use an industry standard color coding that indicates the fuse's current rating.

The industrial style Bussmann ANN fuse is designed to transmit and blow in higher current applications. It typically requires a special fuse holder that bolts the fuse and holder into position. In the RV example in this book, Bussman fuses were used in the electrical circuit between the inverter and the battery bank, and also between the battery bank and chassis ground.

Batteries, Inverters and Load Calculations (Revisited)

In chapter 3, the battery system was fully explained. Now, when using solar power, the concepts remain the same, but you may want to revisit the RV's battery storage capacity and charging system.

One of the justifications for adding solar power is the ability to be independent of and not having to rely too much on shore power hookup. So at this point, you should reexamine your battery system and determine if you want to add more electrical storage capacity. (For calculating your load requirements and expected run times, please refer back to the section titled **"A CASE-IN-POINT EXAMPLE"** found in chapter 3.)

Chapter 3 also covered inverters in great detail. See **DC-to-AC Inverters** in chapter 3. When it comes to adding solar power to your RV and relying on batteries as your source of power, you will need to make sure that the inverter installed in your RV has the capability to also charge batteries. (This is usually an optional feature for inverters.) If not, you should consider purchasing a separate battery charger (see Figure 4.2) so that you have a way to recharge the batteries when solar power is not available or is in limited supply. Since solar recharging is a supplement to your battery charging scheme, it will not always supply 100% of your charging needs. Therefore a conventional battery charger is a very good idea to have as a backup.

Battery charging time is a function of the available voltage because it's the difference in available charging voltage to battery voltage that determines the "speed" at which the battery will get back to full charge. For example, using a good battery charger hooked into shore power will recharge a 50% depleted battery in about 12 hours, whereas the solar charging system will need up to 24 hours of sunlight to bring the batteries back to 100% fully charged. (This translates to 3 or 4 days of good sunlight!) Keeping this in mind and the fact that the last 5-10% of the charge cycle takes almost as long as achieving 90% state-of-charge, it is **highly** recommended that you monitor the discharge level of your battery system and keep them from going into a highly discharged state (50-60% state-of-charge). The best strategy (and one which is recommended by battery manufacturers) for good battery management is to keep the batteries between a 70 and 90% state-of-charge, and to periodically (once a month) bring the battery to a 100% state-of-charge.

Battery charging

There are three stages to the battery charging process – 1) Bulk, 2) Absorption, and 3) Float.

The first stage is **bulk charging**. Current is sent to batteries at the maximum safe rate they will accept until the voltage rises to near full charge level (80-90% state-of-charge). Voltages at this stage typically range from 10.5 volts to 15 volts. There is no correct voltage for bulk charging, but there may be limits on the maximum current that the battery or wiring can take.

The second stage is **absorption charging**. During this stage the voltage remains constant and the current gradually tapers off as the internal resistance of the battery increases during charging. It is during this stage that the charger puts out maximum voltage. Battery voltages at this stage are typically around 14.2 to 15.5 volts.

The third stage is **float charging**. After batteries reach full charge, charging voltage is reduced to a lower level (typically at 12.8 to 13.2 volts) to reduce gassing and prolong battery life. This is sometimes referred to as a trickle charge, since its main purpose is to keep an already charged battery from discharging. Many battery chargers and solar controllers feature PWM (Pulse Width Modulation), which accomplishes the same thing. In PWM, the controller or charger senses tiny voltage drops in the battery and sends very short charging cycles (pulses) to the battery. This may occur several hundred times per minute. It is called "pulse width" because the width of the pulses may vary from a few microseconds to several seconds.

When it comes to the battery **charger** itself, it is important to know the differences between the chargers available on the market. Most consumer, automotive-type battery chargers are bulk charge only, and have little, if any, voltage regulation, but are the most economical to buy. They are fine for a quick boost to low batteries, but are not good to leave on for long periods. Among the more expensive regulated chargers, there are the voltage-regulated ones which keep a constant, regulated voltage on the batteries. If these are set to the correct voltages for your batteries, they will keep the batteries charged without damage. These are often referred to as "taper chargers." What taper charging really means is that as the battery gets charged up, the voltage goes up, so the amps out of the charger go down. They have the ability to charge your battery, but a charger rated at 20 amps may only be supplying 5 amps when the batteries are 80% charged. Top-end chargers now offer "smart" or multi-stage charging. These use a variable voltage to keep the charging amps much more constant for faster charging.

Source: Northern Arizona Wind & Sun. (http://www.windsun.com)

Figure 4.2 Battery charger designed for charging deep-cycle batteries. (Courtesy of Charging Systems International – http://www.dualpro.com)

WIRING DIAGRAMS

Now that you know about each of the components of the solar electrical system, the next phase is to understand how they all get wired together.

I'll start with explaining the typical electrical system in an RV without solar power. This example has an inverter with an optional charger and transfer switch installed (see Figure 4.3). The transfer switch (or relay) automatically switches to battery power if the AC power fails or is turned off. All AC loads connected to the secondary circuit-breaker panel will continue to operate while there is battery power. When the AC power returns, the inverter switches off and the battery charger begins to recharge the batteries. Note that with an inverter that has the transfer switch option, the RV is specifically wired to separate the heavy power appliances, like the refrigerator, air conditioner and heaters, from the system since it is impractical to run them off the inverter. When the relay kicks in, only the appliances and equipment plugged into the secondary distribution panel will receive power. With this type of design, the batteries will always start recharging as soon as you hook up to shore power or turn on the RV generator. Note that if the inverter senses that the batteries have reached a preset low voltage condition, the inverter will shut down and the appliances and equipment plugged into the secondary distribution panel will no longer have power until the batteries are recharged, or you hook up to shore power.

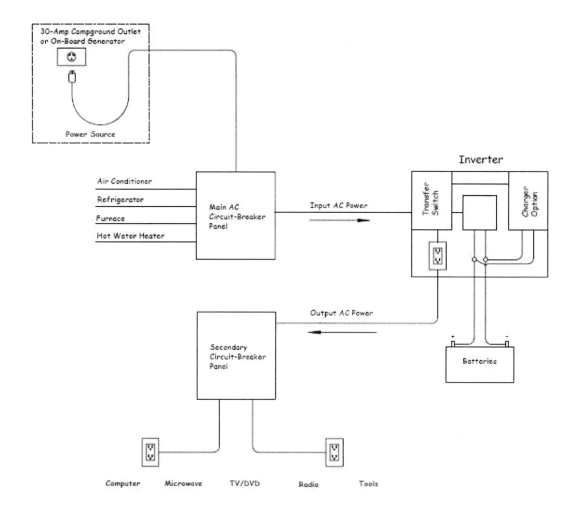

] Figure 4.3 Typical electrical schematic for an RV with a power inverter installed with optional battery charger and transfer switch. The same functionally could be accomplished with a separate relay and charger.

Next, I'll take the example above and add solar power to charge the batteries in addition to the battery charger (see Figure 4.4). I'll also add a DC distribution panel for any DC loads in the system. The design starts again with a power source for the RV. The source could be shore power at a campsite or from an electrical generator installed in the RV. The main AC circuit breaker receives power from the source and distributes it to the larger loads before going into the inverter. The same inverter as above will take care of transferring the power to the secondary circuit breaker panel and charging the batteries when AC power is available. It will also convert the DC power of the batteries to AC power when AC is unavailable.

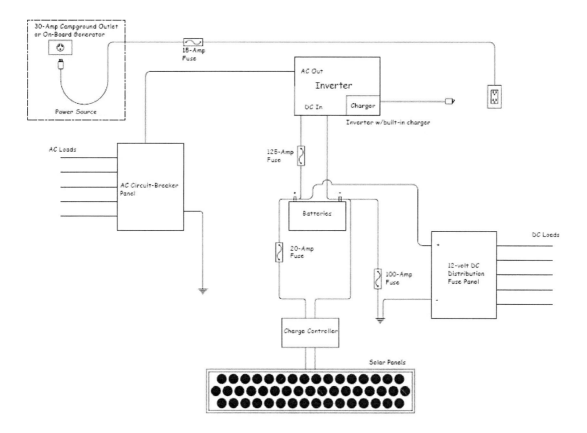

Figure 4.4 Electrical schematic for an RV with the solar power option.

You will notice in this schematic that important fuses have been shown in the circuit to illustrate where over-load protection is applied between the inverter and batteries, solar panels and charge controller, and at system ground. (System ground for an RV is the vehicle chassis.) This same protection less the solar charge controller fuse is part of the typical RV electrical system, but is not shown in Figure 4.2. The size of the fuses is dependent on the inverter and the solar panels installed. The manufacturers of each component will recommend the size of the fuse required.

The addition of solar power requires a charge controller to regulate battery charging, and the solar panels themselves. The panels are connected together, and to the charge controller, with 10 or 12 GA copper wire.

MAINTENANCE

The beauty of solar power is that it is trouble-free. Routine maintenance involves keeping the panels clean and making sure that the connections are clean and solid. To clean the panels, use a product like Windex™ glass cleaner. The panels are covered with tempered glass to protect the photovoltaic cells from weather elements. To achieve maximum energy conversion, keep the panels clean and the connections in good shape.

Chapter 5 – Communications

The mobile office needs the same infrastructure that any normal office requires. If you are using your RV as an office to conduct your business and earn a living, you'll want all the equipment that is used in conjunction with a computer, such as phone, fax, e-mail and Internet access. All of these things are possible in a remote office such as an RV with a little extra effort not associated with a normal office. Clearly, items like Internet access and e-mail are a concern and need to be handled in a special way. This chapter will give you all the details to achieve an office "on-the-road."

PHONE

The phone is still one of the most important items in the home or office. Luckily, cellular phone service has sky-rocketed in the last 10 years and is available just about everywhere. But if you need as much coverage in North America as possible, you should consider boosting your cellular coverage with an external antenna (see Figure 5.1). Boosting your coverage will give you the best chance to always have reception and increase data transfer rates if you use it for fax or dial-up Internet connections. A moderately priced antenna will pick up cellular towers as far as 50 miles from your location.

Phone service is pretty straightforward since most everyone is familiar with cellular phones. For office use, models are available with optional speaker phone plus other options.

Figure 5.1 External cellular antennas. (Courtesy of Wilson Electronics, Inc. – http://www.wilsonelectronics.com)

When selecting a service plan and carrier, you'll want to research phone service providers to determine fees and coverage areas. If you're currently on a local plan, you may want to upgrade to a national plan to avoid roaming charges while you travel.

You may want to consider a "Go Phone" which does not require you to sign up for a contract. Companies such as AT&T and Verizon sell phones that let you purchase blocks of air-time minutes, so you pay-as-you-go. You add more minutes as you need them. This has become such a popular idea, that stores like Wal-Mart sell pre-packaged phones starting at $30 which includes 300 minutes of talk time that can be used anywhere in the country where AT&T has coverage. When you need more air-time, you can purchase pre-paid cards which are sold in 300-600-1000 minute blocks. Again these are available from Wal-Mart or off the Internet from companies such as Net10 (www.Net10.com), where you can purchase minutes on-line for instant activation via a code that you enter into the phone.

Internet History

So what got the Internet started? Believe it or not, it all began with a satellite called Sputnik. In 1957 when the then Soviet Union launched Sputnik, the first man-made satellite, Americans were shocked by the news. The Cold War was at its peak, and the United States and the Soviet Union considered each other enemies. If the Soviet Union could launch a satellite into space, it was possible they could launch a missile at North America.

So in response, President Dwight D. Eisenhower created the Advanced Research Projects Agency (ARPA) in 1958. ARPA's purpose was to give the United States a technological edge over other countries. One important part of ARPA's mission was computer science.

In the 1950s, computers were enormous devices that filled entire rooms. They had a fraction of the power and processing capability you find in a modern computer. Many computers could only read magnetic tape or punch cards, and there was no way to network computers together. ARPA enlisted the help of a company called Bolt, Beranek and Newman (BBN) to create a computer network. The network connected four computers running on four different operating systems. They called the network ARPANET. Without ARPANET, the Internet wouldn't look or behave the way it does today. Although other groups were working on ways to network computers, ARPANET established the protocols used on the Internet today. Furthermore, without ARPANET, it may have taken many more years before anyone tried to find ways to join regional networks together into a larger system.

Source: How Stuff Works (http://computer.howstuffworks.com)

INTERNET

This is one of the most popular and important parts of the book. More and more people are becoming dependent on having an Internet connection for their PC. On-line shopping, weather forecasts, maps, directions and an e-mail connection are just a few examples of things we do on-line almost every day. Just think of how frustrating it would be to not have the ability to look up on the web where the nearest Wal-Mart is, or find out where a branch of your bank is located. It's hard to imagine not having access to all this information. This section will go into detail on how best to increase you ability to stay connected while traveling.

Domain Names, DNSs and URLs

When discussing the Internet, you may hear terms like DNS (Domain Name Service) and URL (Uniform Resource Locator) and wonder what they mean and their purpose in the overall Internet scheme.

First off, most people are familiar with the term **Domain Name,** and know what it's about. It's the unique name companies, organizations and individuals use as the "address" for their web site, e.g. *www.yahoo.com*. Domain names are registered and maintained in a database at InterNIC (part of the U.S. Department of Commerce) and must be renewed each year. Companies like Network Solutions (www.networksolutions.com) offer registration services and on-line forms to help you get your domain registered. The domain name extension is an indicator of the type of site, with the most common being .COM (companies), .ORG (non-profit organizations), .NET (network providers), .MIL (military organizations), .EDU (educational organizations) and .GOV (government organizations). Many more have been added to the list to accommodate the expanding use of the Internet.

Now, it is important to keep in mind that computers and network equipment don't know what *www.yahoo.com* means, for example, because they only work with machine readable information (one's and zero's), so that's where **DNS** fits in. The Domain Name Service is a hierarchical, distributed database that contains mappings of DNS domain names to various types of data, including the unique IP (Internet Protocol) address. DNS enables computers and services to be found by assigning user-friendly names, and it also allows for the discovery of other information stored in the database. So *www.yahoo.com*, for example, is assigned the IP address is 69.147.76.15, which is stored on the DNS servers located on the Internet. When you type in www.yahoo.com in the address bar of your Internet browser, the request is received by one of the DNS servers which translates it into the network address of 69.147.76.15. The networking equipment further translates the network address into the machine readable format of 1010010011011100011100011001101111 and sends it to the Yahoo web server.

The name you type into the address bar of your Internet browser is known as the **URL** address. A URL for a World Wide Web site is preceded with http://, as in the example URL http://www.yahoo.com/. (Browsers automatically add the http:// preface after you type in the www.yahoo.com address and hit enter.) A URL can contain more detail, such as the name of a page of hypertext, usually identified by a file name extension such as .html or .htm, as in http://www.yahoo.com/homepage.html.

As time goes on, the Internet infrastructure continues to expand throughout the country. This means that even while traveling in more remote parts of the country, you'll be able to find a high-speed Internet connection. See sidebar below for more details. In the worst-case scenario, any location that has phone service will provide you with the ability to use a dial-up connection (56 kb/sec), but in reality, this has become a very less-than-desirable method to get a connection, since web page loading, file uploads/downloads, E-Mails with attachments, etc. take a very long time to complete on a slow connection. Most people don't want to wait hours to download files and pictures off the web anymore.

Internet Connections

Internet connections are obtained from the local telecommunication (telecom) companies. The type of connection provided is looked at in terms of bandwidth (connection speed measured in bits/sec), and the technology available based on the telecom's network infrastructure at the location where service is desired. Be aware that the term "High-speed Internet" used in marketing promotions is often misleading, since any service that delivers greater than dial-up connection speeds (56 kbps) is touted as "high speed." In reality, there are many levels of bandwidth available for purchase depending upon the telecom's network infrastructure and how they manage that bandwidth.

The common connection types for residential and small business users are **cable** and **DSL** (Digital Subscriber Line).

Internet cable is often packaged with cable TV since the service is transmitted on the same coaxial cable with the TV signal. Coaxial cable bandwidth is limited to 10 Mbps. Telecoms will divide the available bandwidth between video (TV) and data (Internet) with data typically getting a maximum of 1.5 Mbps. Recently, telecoms are also offering digital telephone service which is often transmitted on the same coaxial cable along with TV and data.

DSL is a more recent technology than cable. With DSL, telecoms use the existing copper telephone lines for data transmission as well as voice. Bandwidth speed is limited to a maximum of 1.5 Mbps and is often sold in fractions thereof (i.e. 768 kbps, 512 kbps, 256 kbps and 128 kbps) at a lower monthly fee. The major disadvantage of DSL is that its availability is limited by the customers distance from the telecom's CO (Central Office): approximately 5 miles.

The following section covers some of the most frequently asked questions about connecting to the Internet. It is assumed that you are traveling with a wireless laptop computer or have a desktop PC with a wireless adapter installed.

Where can I connect?

There are thousands of locations where you'll find Internet access. See sidebars in chapter 2 regarding how Wi-Fi sites are built. Here are generalized categories where you can find "Hot-Spots" to connect:

Campgrounds and RV parks are adding Internet service (Wi-Fi access) at a staggering rate. It is becoming one of the "must have" features to stay competitive and attract business. Parks advertising in publications like Woodall's or parks that have their own web site will most often list Wi-Fi access as a feature. Note that some will make it available for free or for a nominal charge.

Coffee shops like StarBucks™ and Stone Creek Coffee are very popular and serve the dual purpose of getting your java fix and pastry while surfing the web. Again, some will make internet access available for free or for a nominal charge. Many public libraries have wireless onsite that is free of charge to the public. Some may ask for a small donation, which is entirely up to you.

An often-overlooked access point is a rest stop along the Interstate highway system. If Internet access is available, there should be a sign indicating this near the front entrance where maps and weather forecasts are displayed.

Cites such as Chicago, Houston, San Francisco Austin, Champaign-Urbana, Philadelphia, and Long Beach have, or are scheduled to have, city-wide Internet coverage in the near future. Also, there are communities cropping up that install shared, open, wireless service where you may be able to pick up a connection.

If you occasionally stay overnight in a hotel, the major chains have Internet access throughout the hotel.

For a comprehensive list of Wi-Fi Hot-Spots, go on-line to www.jiwire.com. You will find maps and locations, hours of operation, and fees charged where applicable. You will find many sites that are free.

How can I connect?

First, it is presumed that you have Windows networking already configured on your laptop. To get connected to a wireless network, you can use the built in utilities that come with Microsoft® Windows. (For a full detailed description of configuring these services, please see the section in chapter 2 regarding Wireless Networking/Internet Access.) For Windows XP, right-click on the **My Network Places** icon and choose **Properties** from the menu list. Under the **High-Speed Internet** group, select the **Wireless Network Connection**. In the left-hand pane under **Network Tasks** select **"View available wireless networks."** All wireless networks in range will be displayed. Choose the one desired and click connect.

The Windows wireless connection utility is a very basic program without a lot of extra features. You may wish to use the utilities that come with your wireless network adapter to manage connections, since it probably has more useful features.

Is there optional equipment available to increase my ability to connect?

One of the best ways I have found to increase my ability to connect is the addition of a hi-gain, 2.4 GHz omni-directional antenna that is mounted to the roof of the RV with the television antenna. See the discussion of antennas in chapter 2 for complete details. This is particularly handy when staying in a campground that has wireless access, but the signal is weak in certain parts of the campground (dead-spots), and you don't get a connection. With the addition of the antenna, you can pick up the signal from a quarter to a half mile away from the source. The addition of the external antenna also has other benefits. For example, you may wish to get a connection from a Wi-Fi Hotspot (i.e. coffee shop, library, hotel, community center, etc.), but you want to be able to do so from within your RV, rather than unplugging your laptop and carting into the building. With the antenna, you can stop in the parking lot, put up the antenna and get the connection from inside the comfort of your RV.

If you are a traveler to remote areas or always need the ability to connect no matter where you are, satellite dishes are an option for your RV (see Figure 5.2). Companies like Winegard (www.winegard.com/mobile/direcstar.htm) or Satellites In Motion (www.satellitesinmotion.com) sell systems that mount to your RV roof and provide Internet access (as well as satellite TV) - all you need is a clear view of the southern sky. Along with the dish, you will need to subscribe to a satellite Internet Service Provider (ISP) like DirecStar (www.direcstar.com). Although this option is very convenient, it comes with a large price tag. Expect to pay up to $5,000 for the equipment and installation, plus the monthly subscription.

Satellite dish networks

Early satellite TV viewers were explorers of sorts. They used their expensive dishes to discover unique programming that wasn't necessarily intended for mass audiences. The dish and receiving equipment gave viewers the tools to pick up foreign stations, live feeds between different broadcast stations, NASA activities, and a lot of other stuff transmitted using satellites. These broadcasts were free, but viewers had to hunt them down — they didn't get previewed or listed like regular broadcast programming. These signals still exist, and Satellite Orbit magazine publishes a list of today's wild feeds.

Today, most satellite TV customers get their programming through a **direct broadcast satellite** (DBS) provider, such as DirecTV or DISH Network. The provider selects programs and broadcasts them to subscribers as a set package. Basically, the provider's goal is to bring dozens or even hundreds of channels to your TV in a form that approximates the competition, cable TV. Unlike earlier programming, the provider's broadcast is completely **digital**, which means it has much better picture and sound quality. Early satellite television was broadcast in **C-band radio** — radio in the 3.7-GHz to 6.4-GHz frequency range. Digital broadcast satellite transmits programming in the **Ku frequency range** (11.7 GHz to 14.5 GHz).

There are five major components involved in a **direct to home** (DTH) or **direct broadcasting** (DBS) satellite system: the programming source, the broadcast center, the satellite, the satellite dish and the receiver.

- **Programming sources** are simply the channels that provide programming for broadcast. The provider doesn't create original programming itself; it pays other companies (HBO, for example, or ESPN) for the right to broadcast their content via satellite. In this way, the provider is kind of like a broker between you and the actual programming sources. (Cable TV companies work on the same principle.)
- The **broadcast center** is the central hub of the system. At the broadcast center, the TV provider receives signals from various programming sources and beams a broadcast signal to satellites in geosynchronous orbit.

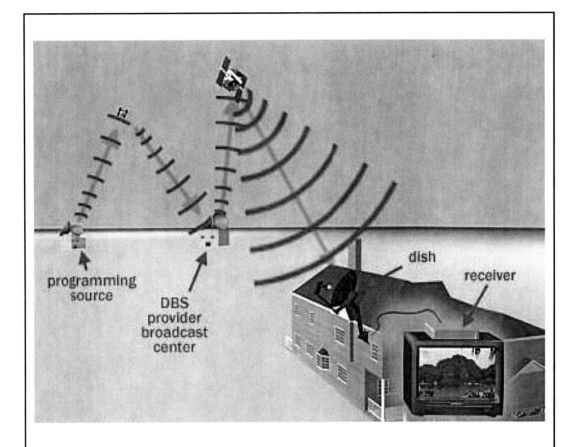

- The **satellites** receive the signals from the broadcast station and rebroadcast them to Earth.
- The viewer's **dish** picks up the signal from the satellite (or multiple satellites in the same part of the sky) and passes it on to the receiver in the viewer's house.
- The **receiver** processes the signal and passes it on to a standard TV.

Source: How Stuff Works (http://computer.howstuffworks.com)

Figure 5.2 Satellite dish systems. (Courtesy of Winegard – http://www.winegard.com)

Are there other items I may need to stay connected?

Yes, there are miscellaneous items you may want to keep in your PC "toolkit." A few very handy items I keep on hand are:

- A 10 ft. long network patch cable for Internet connections that are only available through a hard-wired connection to your laptop's network port.
- A 10 ft. long USB A-B network cable for USB type connections.
- A 10 ft. long USB extension cable for those hard to reach places.
- A USB Flash Drive to store backups of important documents and files. Also very handy if you need to transfer files to another laptop or PC.
- A 10 ft. long RJ-11 telephone cord for dial-up connections.

All these items are available from stores like Best Buy, Office Depot, Office Max or Computer City.

How do I connect to the Internet with a dial-up connection?

If this is your only option, here are instructions to make a dial-up connection.

First, you will need to sign-up for an Internet Service Provider (ISP) that sells dial-up accounts. If you are traveling throughout the country, you'll want a provider that can give you a toll-free 1-800 number for the connection so you don't pay for long distance charges along with your ISP monthly service plan. Check out Yahoo! DSL (www.yahoo.com) or Net Zero (www.netzero.net) for example.

Next, follow these setup instructions (These instructions refer to the procedures for Windows XP and assume your computer has a modem that has been setup. See below for modem setup.)

The process is done through a Windows setup wizard. To access the wizard, open the Control Panel and select Network Connections from the list. When it opens you will see:

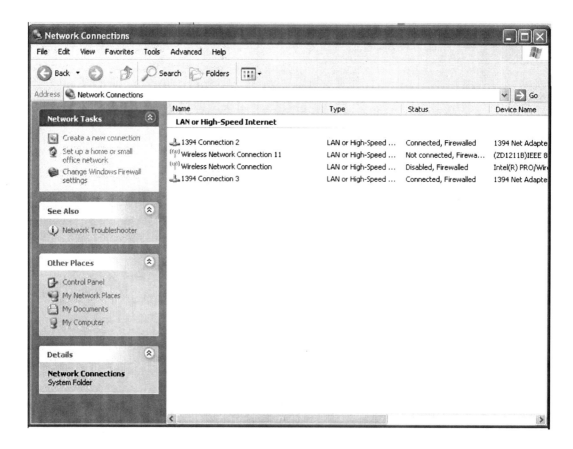

Under Network Tasks, click on Create a new connection.

Click Next to continue. On the following screen, select Connect to the Internet and click Next to continue.

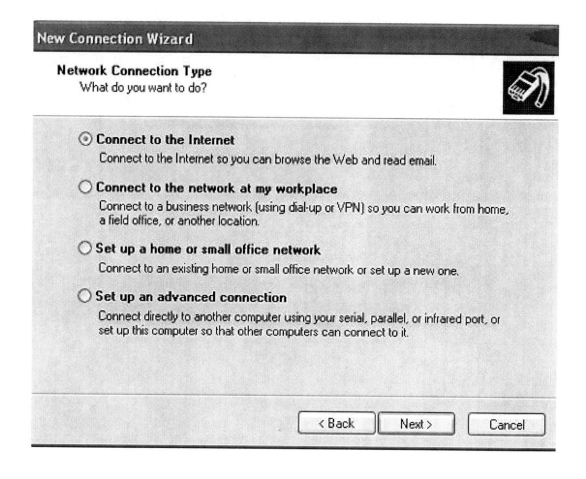

On the next screen, select Set up my connection manually and click Next.

From the next screen, select Connect using a dial-up modem. Click Next.

New Connection Wizard

Internet Connection
How do you want to connect to the Internet?

- ⦿ **Connect using a dial-up modem**
  This type of connection uses a modem and a regular or ISDN phone line.

- ○ **Connect using a broadband connection that requires a user name and password**
  This is a high-speed connection using either a DSL or cable modem. Your ISP may refer to this type of connection as PPPoE.

- ○ **Connect using a broadband connection that is always on**
  This is a high-speed connection using either a cable modem, DSL or LAN connection. It is always active, and doesn't require you to sign in.

[< Back] [Next >] [Cancel]

The next screen asks you to give the connection a name. Type anything you want for the name. Click Next.

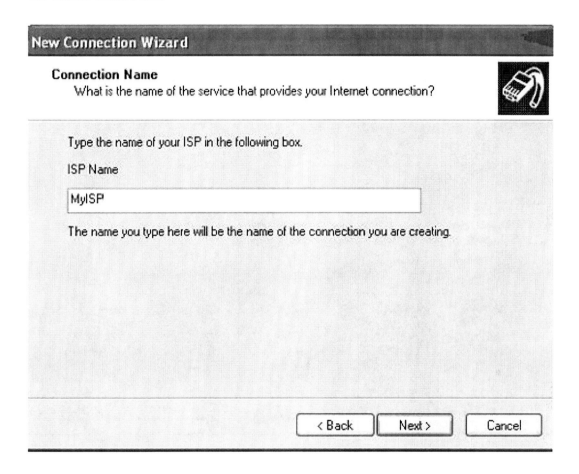

Enter the phone number and click Next.

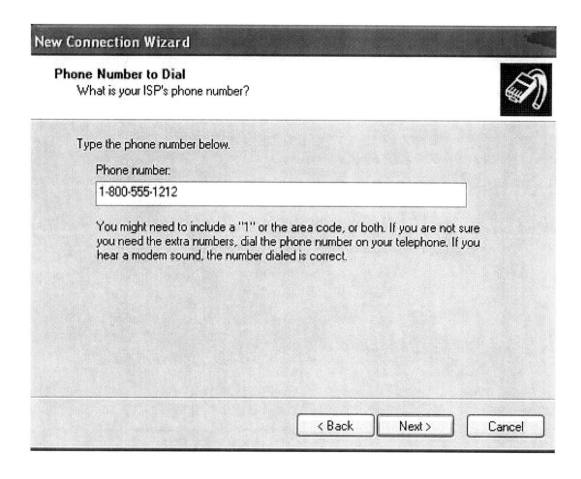

On the next screen, enter your account information. This will be provided by the ISP. Leave the two check boxes marked. Click Next when done.

Finally, the following dialog will appear. Make sure to check the "Add a shortcut to this connection to my desktop." Click Finish.

You will now have a shortcut on the desktop for making the dial-up connection (in this example My ISP).

When you click on the dial-up icon (My ISP in this example), a dialog box will appear, ready to make the connection. The telephone number and your user name will be filled in. Provide your password and click Connect.

To install a modem

You must be logged on as an administrator or a member of the Administrators group in order to complete this procedure. If your computer is connected to a network, network policy settings might also prevent you from completing this procedure.

1. Open Phone and Modem Options in the Control Panel. If you are prompted for location information, specify the dialing information for your location, and then click **OK**.
2. On the **Modems** tab, click **Add**.
3. Follow the instructions in the Install New Modem Wizard.

Notes:
1. To open Phone and Modem Options, click **Start**, click **Control Panel**, and then double-click **Phone and Modem Options**.
2. If the Install New Modem Wizard does not detect your modem, or you cannot find it listed, click **Related Topics** for instructions about installing an unsupported modem.
3. If you are installing an external modem, and a cable is not provided, refer to the manufacturer's instructions for cable requirements. Most common cables will work, but some cables do not have all of the pins connected. Do not use the 9-to-25 pin converters that come with most mouse hardware, because some of them do not carry modem signals.

To configure a modem for dial-up connections:
1. Open Network Connections.
2. Click the dial-up connection you want to configure, and then, under **Network Tasks**, click **Change settings of this connection**.
3. On the **General** tab, under **Connect using**, click the modem you want to configure, and then click **Configure**.
4. Under **Hardware Features**, select the check box options you want to enable.
5. If you want to enable the modem speaker, select the **Enable modem speaker** check box.

Notes:
- To open Network Connections, click **Start**, click **Control Panel**, and then double-click **Network Connections**.
- To ensure compatibility, you need to use the same kind of modem as the one connected to the remote access server, select the same initial speed, and enable the same features. If you do not select the same model, at least select a modem with the same ITU-T standard as the modem on the server.
- Selecting a feature that is not supported by your modem has no effect on its performance.
- Network Connections automatically configures connections according to the devices that are available. For example, you might use a laptop and docking station when you are in the office, with your connection configured to use the docking station modem. When you are on the road, and the docking station modem is not available, the connection is automatically configured to use the laptop's PCMCIA modem. When you return to the office and re-dock your laptop into its docking station, Network Connections detects that the docking station modem is available again, and automatically reconfigures the connection to use the docking station modem.

Use Windows Help & Support for more topics. Search for **Dialup Connections**

When you want to connect, hook up your laptop through its phone jack to the RV's phone jack using a telephone cable (Type RJ-11). Then click on the shortcut icon that was put on your laptop's desktop during the setup of the dial-up connection. Your computer should start dialing the number and show when it is connected. Open up your web browser (i.e. Internet Explorer) and surf as usual or send mail.

A note about dial-up connections: as was pointed out before, the maximum connection speed is 56 kb/sec, and therefore it can take 5-10 times longer to open a web page or file as compared to a high-speed connection. Be patient! Don't interrupt the computer while it is working on loading pages or transferring files.

What are some popular links that may be helpful while traveling?

Here are some popular links I've put together over time. Copy and paste the http address and press enter. This is not an exhaustive list by any means, so when in doubt – Google™ It!

For Helpful Links, see Appendix C

E-MAIL

This section goes into a detailed discussion on setting up e-mail accounts (multiple accounts if necessary) and how to handle issues with sending and receiving email while you travel.

Many people get by with a simple account through providers like Yahoo!™, MSN Hot Mail™, or AOL™ that are accessible with just an Internet connection and a web browser. These are fine if you don't need the advanced features that come with email viewers like Microsoft® Outlook or Microsoft® Express. For those of you that do, please read on!

The following is an example of how to setup an Outlook 2007 or Outlook 2003 e-mail account. (This example also applies to Outlook versions 2000 and 2002.) The mailbox for this account resides at the company that provides your Internet access when you are at home. For this example, Time Warner Cable is the ISP for your Internet. When you signed up for the service, you received a POP3 (the most widely used by ISPs) type mailbox for receiving your mail. Before getting started with setting up the e-mail, you will need at a minimum the following pieces of information as supplied by the ISP:

> **E-mail address**: JDoe@twcable.com
> **POP3 server address**: mail.twcable.com
> **SMTP server address**: smtp.twcable.com
> **User name**: jdoe1234
> **Password**: xc6tw7z

Internet E-Mail

One nice advantage of the Internet is e-mail correspondence. It's hard to image not having the convenience of sending and receiving messages, files and pictures with the expediency of the e-mail system.

The two most popular ways to use e-mail is either via a Web browser, *e.g. Internet Explorer,* or through a special e-mail messaging program that runs on your computer, *e.g. Microsoft Outlook.*

Browser-based e-mail has many benefits. Your e-mail delivery, user interface, administration and storage of messages are all managed by the e-mail host, *e.g. MSN Hotmail.* Often the service is free or bundled with another product. Being browser-based allows you to access your mail anywhere you can get an Internet connection and there is no setup to worry about. The disadvantage is that browser based e-mail programs do not have many of the advanced features of a Windows based e-mail messaging program. The most apparent are formatting of the message (the message editor), the need to download attachments before being able to view them, limited message storage space and managing junk mail to name a few. They also lack good contact management and calendar functions.

Windows messaging programs can offer many more features, like being able to use your favorite text editor, creating custom group lists, delivery confirmation and appointment scheduling. The big drawback is in getting the email program setup to send and receive the mail. No matter how many setup wizards programmers devise, you still need to have a basic understanding of what POP3 and SMTP servers and Internet addresses are all about, where to find the information through the e-mail service provider and where the information needs to go in the program. Another problem for the traveler is the issue of sending e-mail. ISPs limit the sending of email to only their customers. So when you leave home base and try to send e-mail from a remote location (in other words a different ISP), the mail will not be delivered. Therefore if you travel a lot, it is advantageous to subscribe to a service that (for a fee) will let you send e-mail from their server, no matter where you are located. (Check out www.authsmtp.com as an example.)

E-Mail Servers

For those of you interested in more detail about e-mail delivery on the Internet, this sidebar describes the key elements in the delivery system.

The Internet e-mail system has many similarities with the traditional mail delivery system. The Internet has post offices, addresses, mailboxes, delivery services, receipts, forwarding, and return mail.

The post offices are dedicated computers that are connected to the Internet and run software to send and receive e-mail messages. An electronic post office performs the same functions as any city post office, but instead of serving a city, the electronic post office serves a specific domain group (*e.g. AOL.com or IBM.com*). A domain is made up of a group of people who are members of the domain and have mailboxes in that domain similar to people who are part of a community and have an address and mailbox in that community. Each person in the domain can have one or more email address (*e.g. BobSmith@IBM.com, BSmithPrivate@IBM.com*), where their mail is delivered. Incoming mail destined for a member in the domain is sent to the domain's POP3 (Post Office Protocol 3) server, which processes all incoming mail and delivers it to the appropriate mailbox. Outgoing mail from a member of the domain is processed by the SMTP (Simple Mail Transfer Protocol) server. The SMTP protocol governs the exchange of electronic mail between message transfer agents. When setting up e-mail programs, the internet address for the incoming POP3 mail server (*e.g. mail.aol.com*), the outgoing SMTP server (*e.g. mail.authsmtp.com*), the mailbox address (*e.g. BobSmith@IBM.com*) and the logon/password for the e-mail account are the essential pieces of information required.

Other features of interest are the ability to get a delivery confirmation for received e-mail and notification for undeliverable mail. Undeliverable mail can occur if the e-mail address is not valid or the receiving e-mail server is unavailable. Post offices will usually attempt to deliver the mail three times over a 24-hour period and will send notification to the sender if it fails after the third attempt.

Assuming now that Outlook is installed on you laptop, follow these step-by-step instructions:

Start by opening the Control Panel (in Classic View) on your laptop. Click on the Mail applet and choose E-mail Accounts…

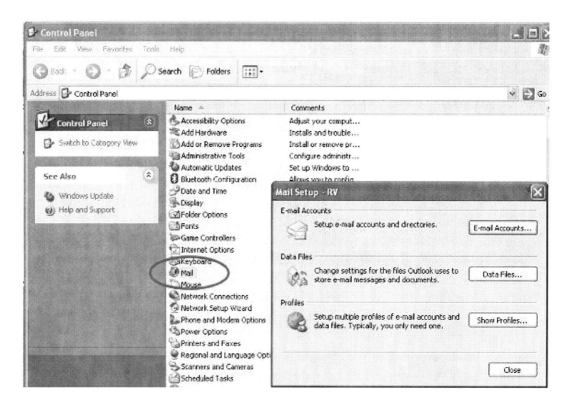

The E-mail Accounts dialog box will appear. Select "Add a new e-mail account" and click Next.

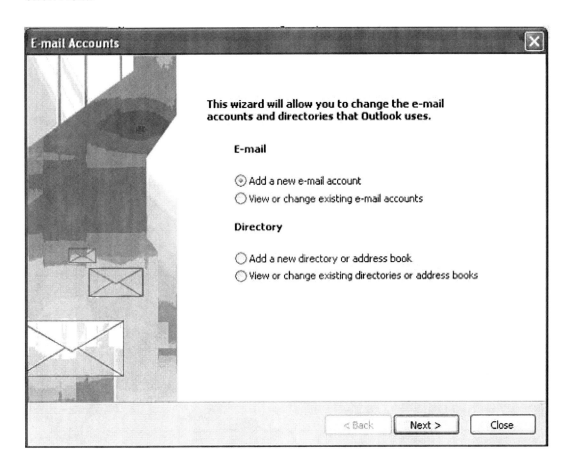

The Server Type dialog will open. Choose **POP3** and click Next.

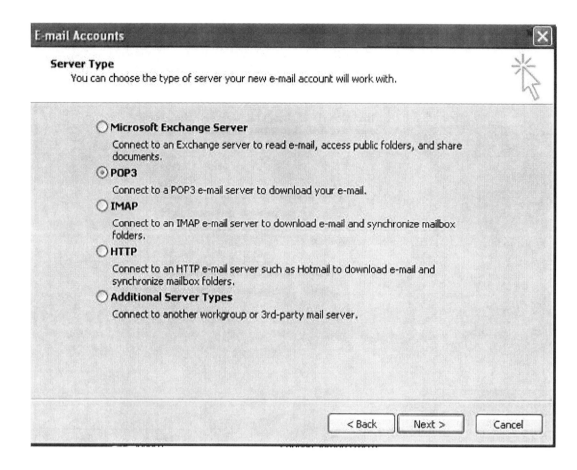

Enter the Internet E-Mail settings (POP3) as provided by your ISP and click Next.

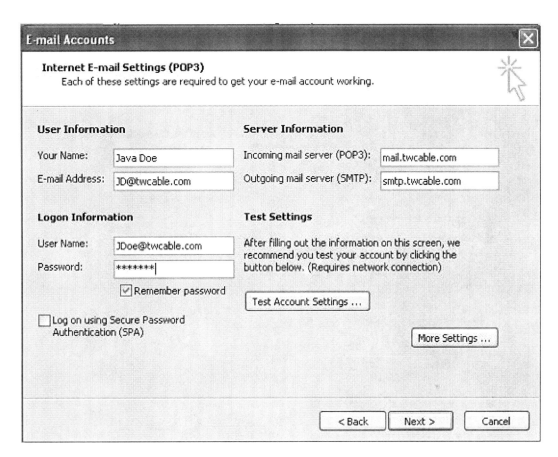

(Note: The Your Name information as entered above will be what recipients see displayed in their inbox when you send them mail.)

Next, click Finish.

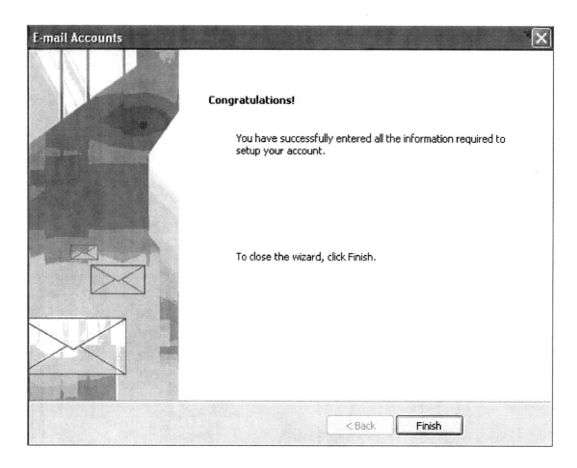

If you have multiple POP3 e-mail accounts for different e-mail addresses, you can turn on e-mail profiling in Outlook. To do so, open the Control Panel, click on Mail, then on **Show Profiles**. Next, click on the **Add** button and proceed to enter in the next e-mail account, as you did above. After the account is added, be sure to check the box on the Mail dialog screen that says "Prompt for a profile to be used," so that when Outlook starts up each time, you will get an opportunity to choose the mailbox you want.

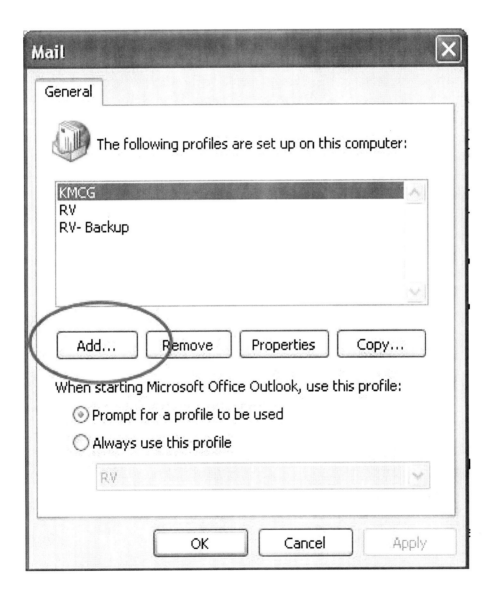

The Outlook Data File

Outlook stores all your e-mail in one file for each e-mail account that you setup. The file has all the email located in your Inbox, Sent Items, Deleted and Drafts mail folders. When you create new mail folders in your mailbox to organize and group the emails you wish to save, these are also stored in that file (see Figure 5.3). By default, Outlook creates a file called **Outlook.pst** buried deep within your personal Documents and Settings folder (which is only accessible by you). In the example setup above, the file for Java Doe's e-mail is located in **C:\Documents and Settings\JDoe\Local Settings\Application Data\Microsoft\Outlook\Outlook.pst.** Note that if you create another email account, the data file will have a different name. Outlook appends a number to the end of the file name for each new mail file. Thus, the next Outlook data file will be called **Outlook1.pst.** Now you may be wondering why this is handy to know. If you are like me, the e-mail I save is very important, and I like to make a backup of it every so often. To do this, I need to know the name of the file and where it is located on the computer. I typically back it up once a week to my USB flash drive.

Working and Living Independently on the Road 153

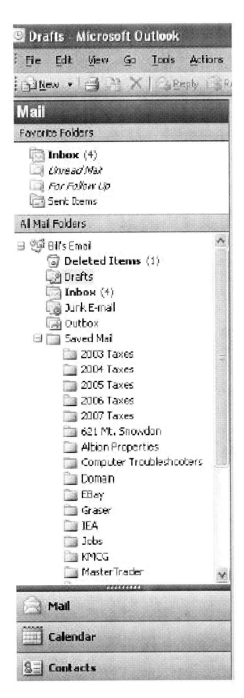

Figure 5.3 Example of saving your e-mail in organized folders. Create new folders in your mailbox by right-clicking on the top-level folder and choosing "New Folder."

You may want to reorganize where your data file(s) are located to simplify backup and possibly give the files more meaningful names. Here is how you manage the data file.

The Outlook data file is managed with the Mail applet found in the Control Panel (please use the Control Panel's Classic View to follow along). In the example below, the e-mail account was setup earlier. Open the Control Panel, double click on Mail, then click on Show Profiles.

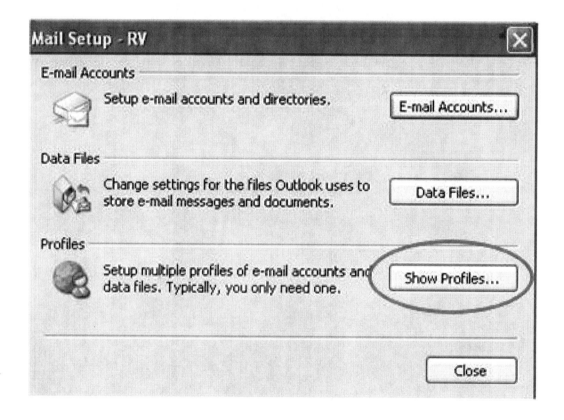

Next, select the profile called Java Doe for this example, then click on Properties.

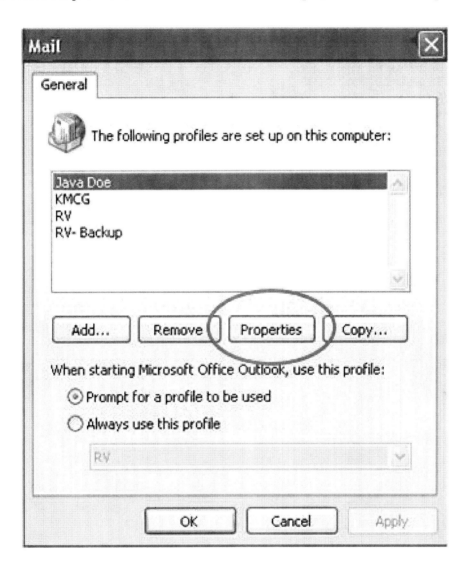

When the mail setup for Java Doe appears, click on Data Files.

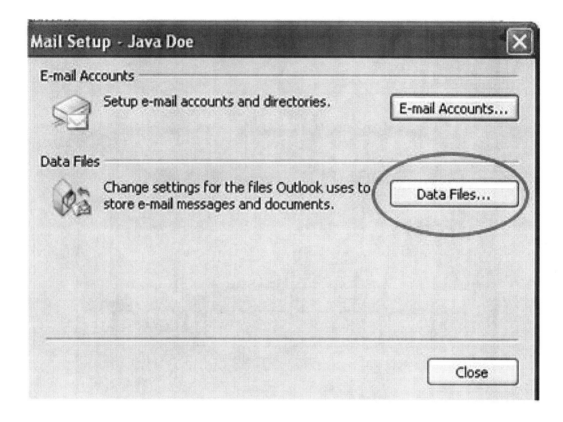

The Outlook Data Files for Java Doe will open.

In this example, Java Doe's Outlook email account has not been used yet and contains no mail. In the following steps we are going to create a new Outlook data file with a more descriptive name and store it in an easier to find location on the computer. You will note that there is an Open Folder button that lets your choose an existing data file if you already have one you would like to use on the computer. In this case we don't, so click on Add to proceed.

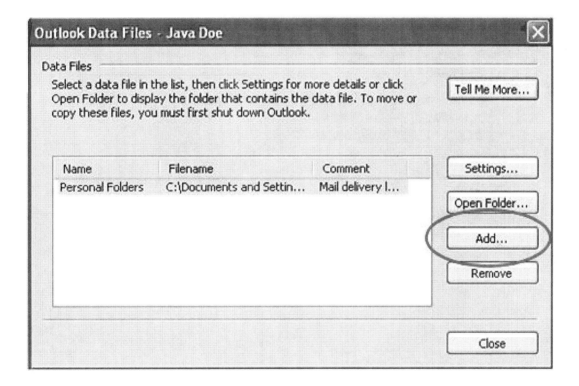

A dialog box is presented that lets you add a new Outlook Data File. Since this example is for Outlook 2003, keep the type of storage set to Office Outlook Personal Folders File (.pst), and then click OK.

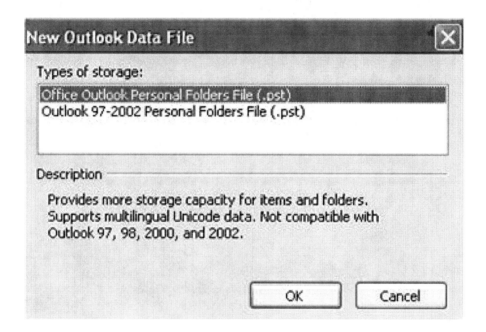

You will see a dialog that lets you create the new data file. In this case, I have previously created a folder on my C-Drive called **Email File**s where I like to store my email. Select this location in the **Save In** dropdown box.

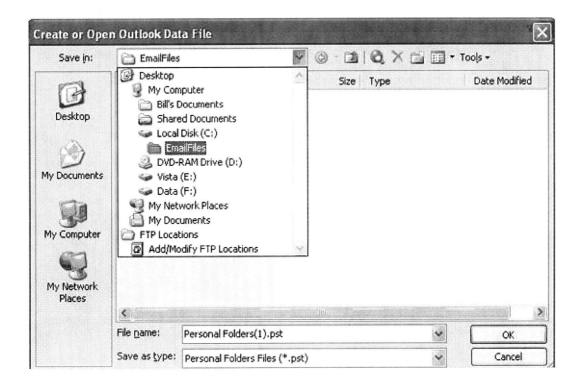

Next, enter in a file name in the **File name** box. In this case I entered JDoeEmail.pst and then clicked OK.

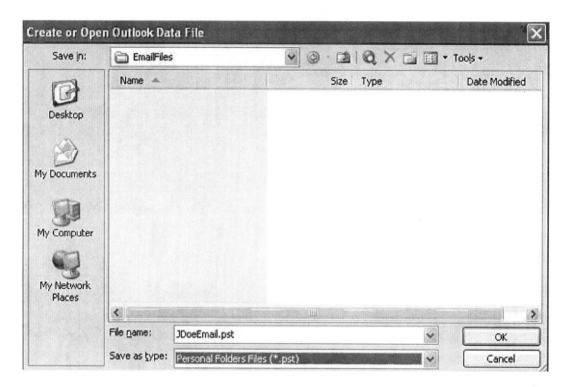

The next screen will show the Personal Folder information for Java Doe Email. Click on OK to continue.

This will bring you back to the Outlook Data Files screen for Java Doe showing the new data file along with the original data file created when Outlook was first setup (currently empty).

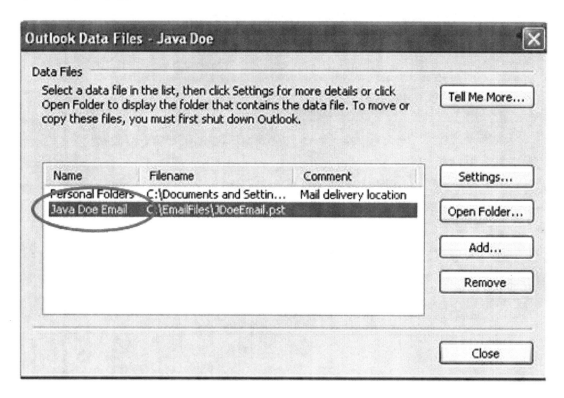

The Comment column will show where the mail will be delivered. In this case it is pointing to the original location. We want to point it to the new location (C:\EmailFiles\JDoeEmail.pst), so click on the Close button to bring you back to the Mail Setup screen. Click on the E-mail Accounts button to continue.

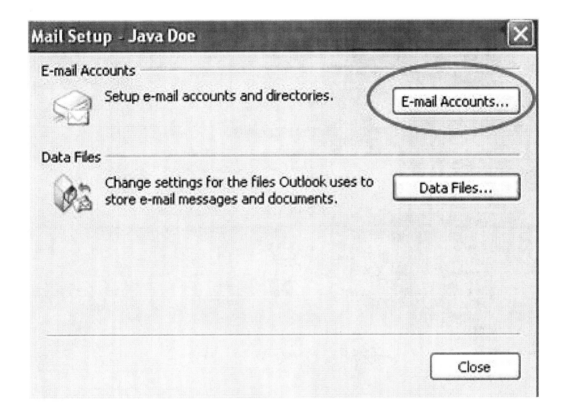

Notice on the E-mail Accounts screen that "View or change existing e-mail accounts" is automatically selected, since accounts exist. Click Next to continue.

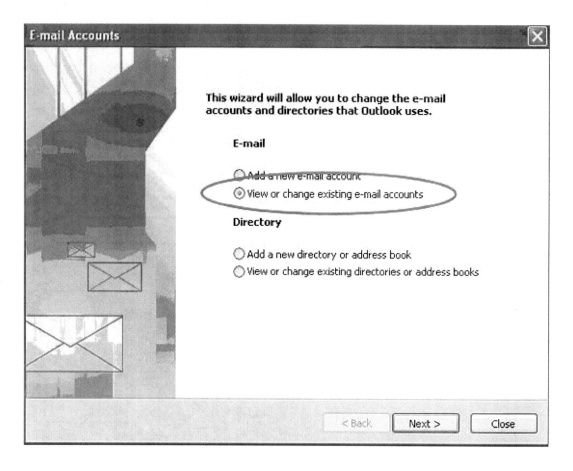

Next, we want to change the mail delivery location. In the dropdown box under "Deliver new e-mail to the following location:" select **Java Doe Email** from the list, then click the Finish button.

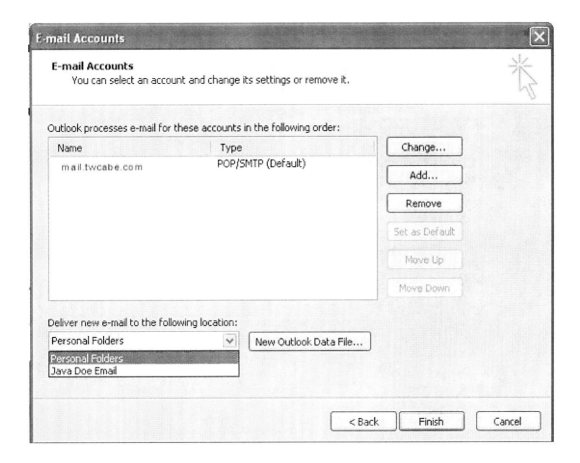

You will see a warning that tells you that the location has changed. Click OK to accept.

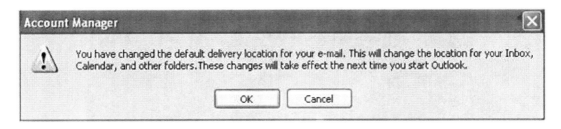

This brings you back to Mail Setup. If you click on the Data Files button, you'll now see that the delivery location is Java Doe Email.

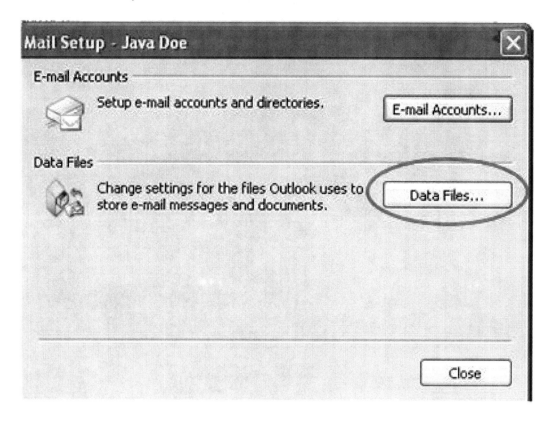

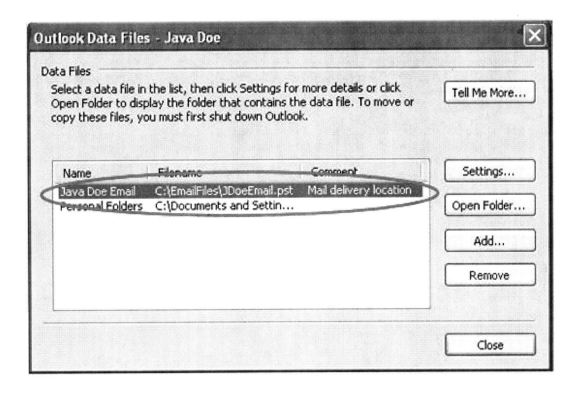

I also delete the original "Personal Folders" location since it not needed and may cause confusion later when viewing the settings. To delete it, click on that line so it gets highlighted then click the Remove button. Outlook will display a warning. Click Yes to accept.

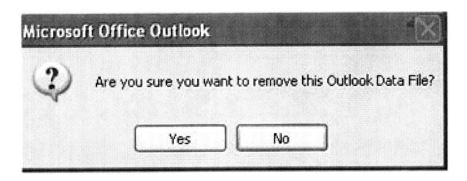

The Outlook Data Files screen will now look like this. Click Close.

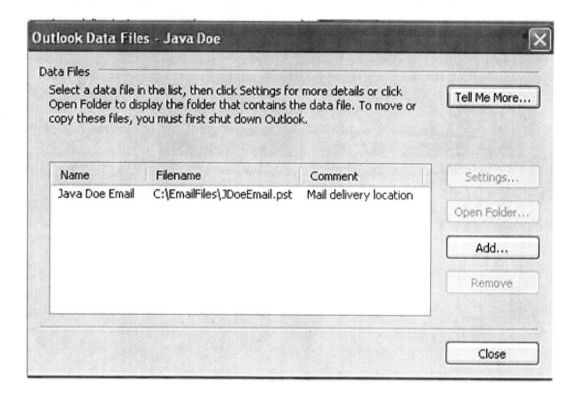

Click Close on the next screen.

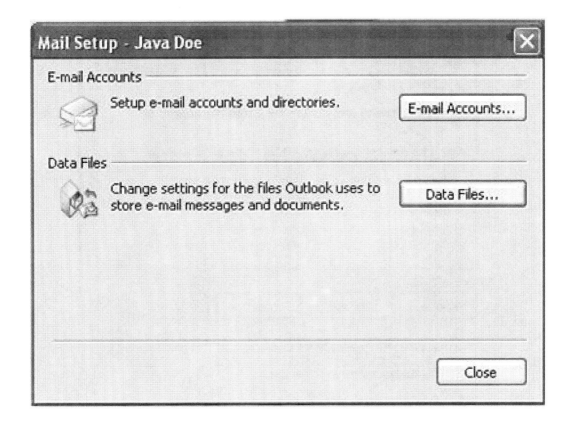

Finally, click OK to exit the Mail applet.

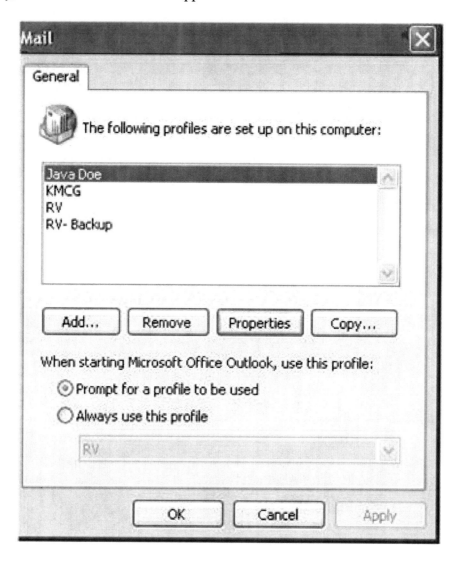

Advanced Features of Outlook Setup

You should be aware of a few setup options that may be required when setting up your e-mail account. These features can be selected after you have your e-mail setup. The following is a list of steps that show how to access these options. In this example, the e-mail account was previously setup and now you're going to make modifications.

The Outlook advanced setup features are managed with the Mail applet found in the Control Panel (please use the Control Panel's Classic View to follow along). In this example, the e-mail account was setup earlier. Open the Control Panel, double click on Mail, then click on Show Profiles.

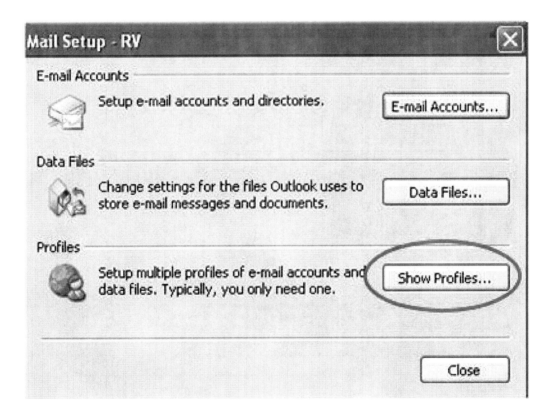

Next, select the profile called Java Doe for this example, then click on Properties.

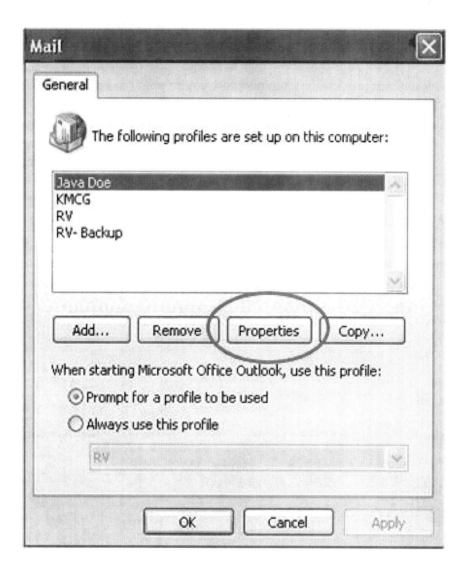

Working and Living Independently on the Road   173

When the Mail Setup for Java Doe appears, click on E-mail Accounts to continue.

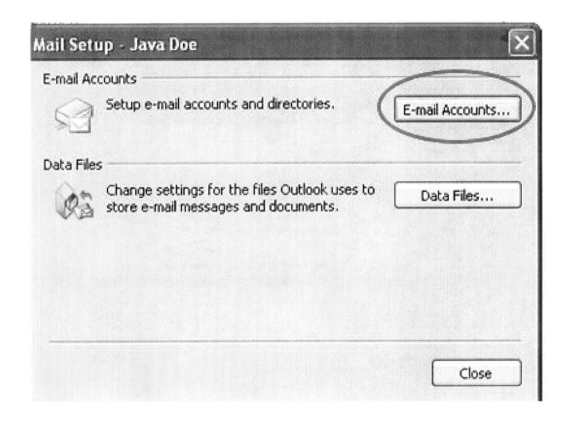

Notice on the E-mail Accounts screen that "View or change existing e-mail accounts" is automatically selected since accounts exist. Click Next to continue.

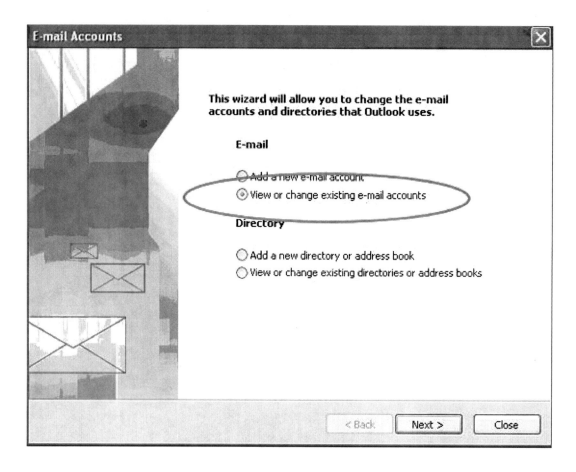

The E-mail Accounts dialog box appears. Click on the Change button to continue.

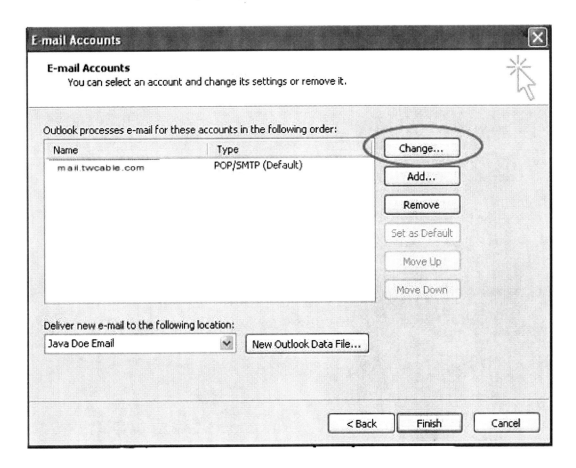

The Internet E-Mail Settings opens. Click on the More Settings button.

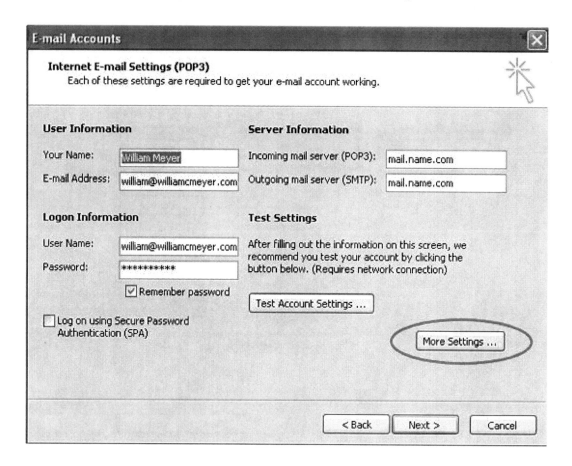

This will bring up the options for E-mail Settings.

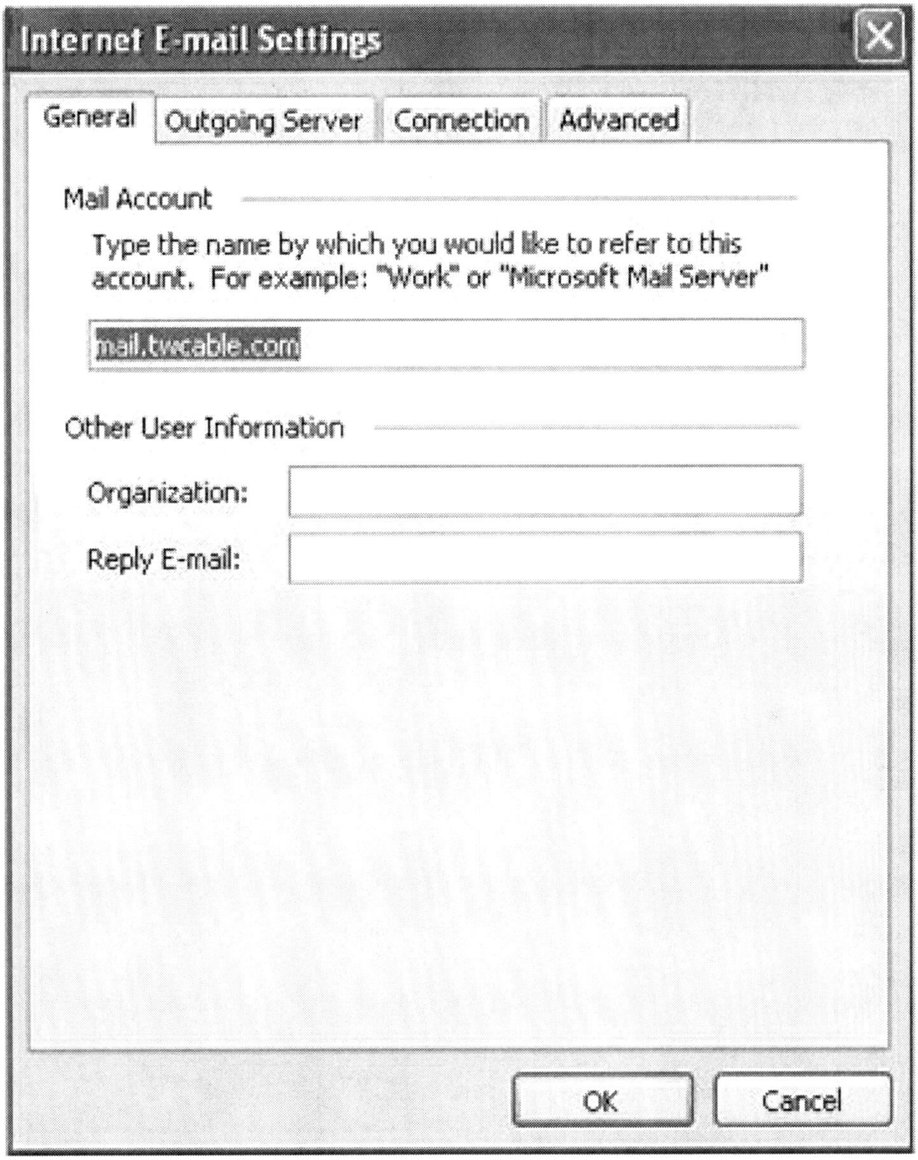

Click on the **Outgoing Server** tab to bring up this screen.

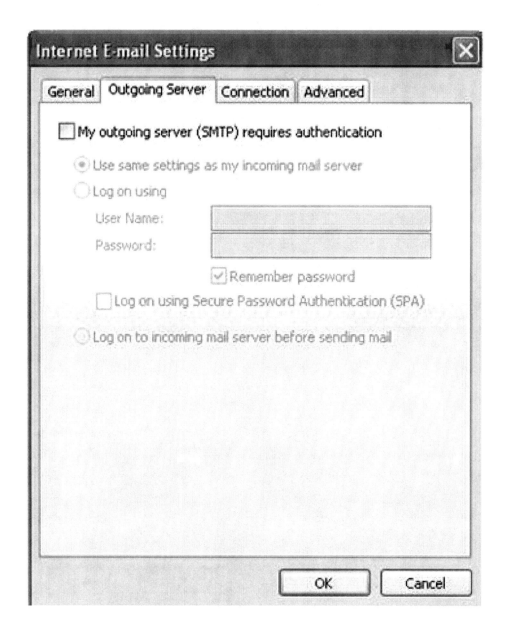

When the **Outgoing Server** tab appears, nothing is selected. Click on the "My outgoing server (SMTP) requires authentication" checkbox. This makes the options available.

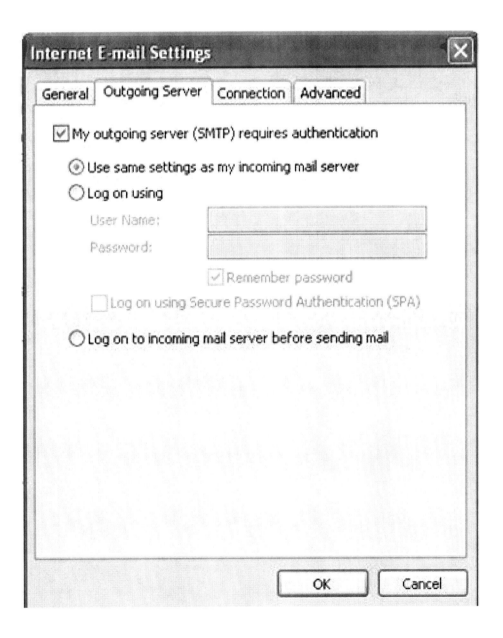

Your ISP will tell you if their mail delivery server (Outgoing SMTP server) requires you to login first before sending your mail. If so, the login will either be the same as for the Incoming (POP3) server in which case you then leave the settings as shown above, or they will provide a different User Name and Password. If so, choose "Log on using" and fill in the User Name and Password. (You will not be asked to enter this each time you send mail, since Outlook uses what you stored on this tab.)

Some times ISPs change what are called the "Server Port Numbers" for security reasons. If so, the ISP will provide you with those numbers, which you edit as needed. (The default values are shown below and are the most widely used.) Click the **Advanced** tab to see those options.

Edit the Server Port Numbers as needed. When you are finished with your changes, click OK.

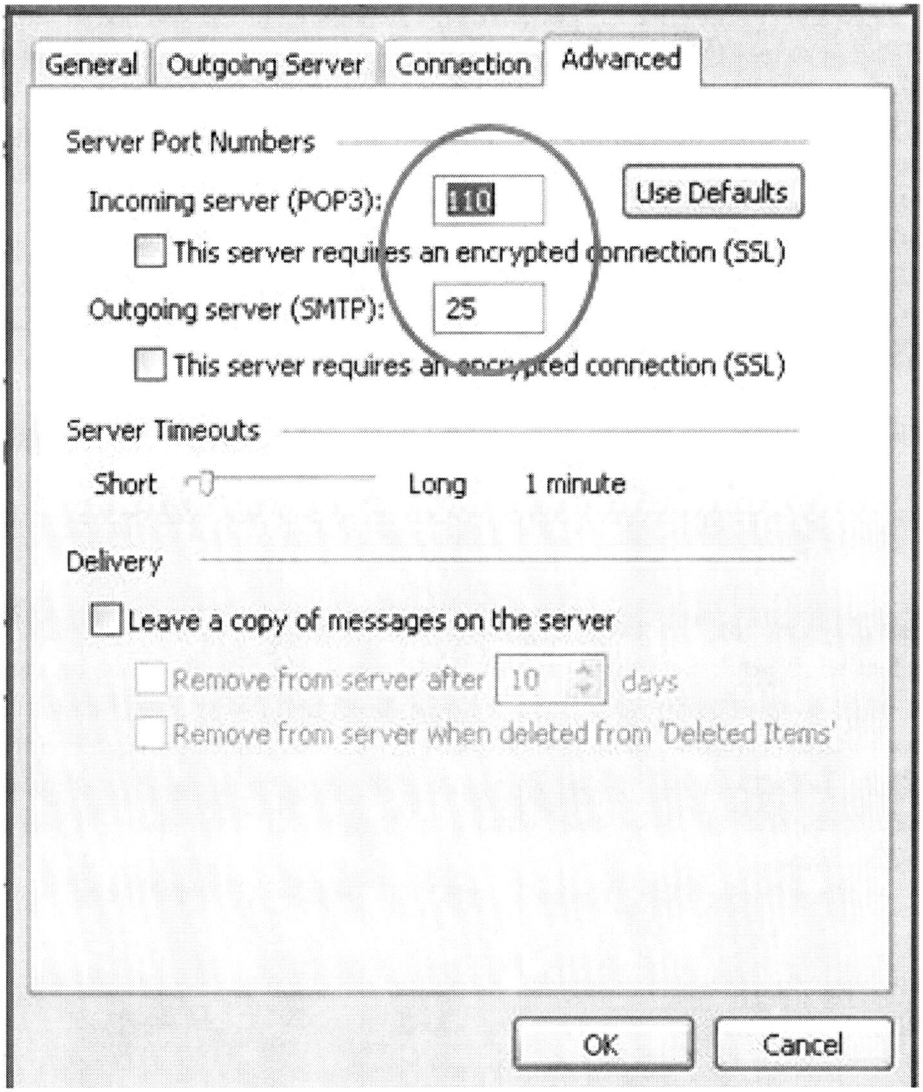
General
Outgoing Server
Connection
Advanced
Server Port Numbers
Incoming server (POP3):
110
Use Defaults
This server requires an encrypted connection (SSL)
Outgoing server (SMTP):
25
This server requires an encrypted connection (SSL)
Server Timeouts
Short
Long
1 minute
Delivery
Leave a copy of messages on the server
Remove from server after
10
days
Remove from server when deleted from 'Deleted Items'
OK
Cancel

There is a **Delivery** section at the bottom of the **Advanced** tab. If you use a Palm Pilot, Blackberry or other PDA to receive e-mail, it is a good idea to select the "Leave a copy of messages on the server" checkbox, so that you can see the incoming mail on the device and in Outlook after the mail is opened. Otherwise, if this is not selected, when you open the e-mail, it is copied to the Outlook Inbox (or PDA Inbox depending on where it was opened first) and removed from the ISP's server. By not removing it from the server, this gives you the opportunity to read the e-mail using your PDA and then later filing it away in your Saved folders in Outlook. (**Reminder** – Folders you create in Outlook are only located on your computer. You will not see them on your PDA and they are not available on the ISP's server.)

Also check the "Remove from server after X-days" and set the number of days. 10 is a good starting place. You don't want to set the value too high since your mailbox will fill up at the ISP and may reach its capacity, at which time you will no longer receive mail.

I suggest checking the "Remove from server when deleted from Deleted Items." This means that when you delete mail from Outlook, it will not take up space on the ISP's server. **Important** – When you delete an e-mail in your Inbox or Sent Items folder, it is moved to the Deleted Items folder but it is not completely removed until you "Empty the Deleted Items" folder. To permanently remove items, right-click on the Deleted Items folder in Outlook and choose the **Empty "Deleted Items" Folder**. See screenshot below.

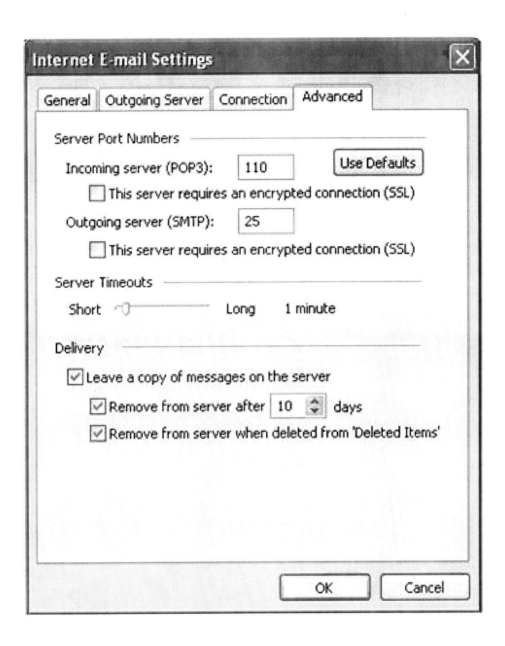

Sending E-Mail While Traveling

There are pit-falls with sending e-mail while traveling away from your home base. This has come about as a direct result of e-mail spamming and the abhorrent amount of junk e-mail delivered to our mailboxes. One of the measures that ISPs take is to not allow users from other computer domains to send out email through their outgoing (SMTP) server. In the example above, Java Doe sends email through the **twcable.com** domain. This works fine until he leaves home base and logs into an ISP provider that uses the domain **MyDifferentDomain.com** for sending e-mail. Since Java is from the **twcable.com** domain, his outgoing e-mails will be rejected and sent back with an error message when trying to use a different domain and SMTP server. The incoming e-mail will still be received since it is delivered to the POP3 server in the **twcable.com** domain.

There is a solution though. Companies, *e.g. www.authsmtp.com*, sell a service that allow you to purchase the use of their SMTP outgoing e-mail server from any domain. For a small fee based on the volume of email you send, you will be assigned an Outgoing e-mail (SMTP) server address which you put into your Outlook E-Mail Account setup. You will also receive a User Name and Password to log in to their server. (This gets setup on the Outgoing Server tab as discussed above under the Advanced Features of the Outlook Setup section.) You will then be able to send e-mail from anywhere, including your home base.

You are now ready to receive and send e-mail when connected to the Internet. (Remember, you must be connected first!)

Chapter 6 - An Illustrative Example

I've put together this illustration of how I setup my conversion van to support my business and lifestyle while traveling all over the country. I offer computer services, web site design, and do contract work as a computer programmer. This has been an ideal situation since I can pick up work almost anywhere, and programming can be done where ever I feel like doing it! I do this in a 19 ft. Home & Park™ RoadTrek 190 motor home.

Home & Park™ RoadTrek 190 motor home.

As your peruse this section, you will notice that I have taken extra care to supply myself with "free" electricity to power my computer equipment and other electrical needs. I like the freedom of "boondocking" and traveling to remote parts of the country. This requires more electrical storage capacity than you find in the average RV. I have sized the system to allow a minimum of 3 days between recharging of the batteries (4 deep-cycle 100 amp-hour house batteries). I have yet to run into a situation where I don't have power to supply my needs having 4 batteries and 130 watts of solar power for recharging. In a pinch, I have an on-board 2800-watt generator that can recharge batteries and supply power for my equipment, microwave oven, refrigerator and A/C.

Let's take a look:

I spent a lot of time designing the layout of my "office space" to achieve a fully functional work environment. Since I'm in a small Class-B motor home, I've had to make some spaces pull double duty for other functions. Those of you traveling in Class-A and Class-C motor homes can probably dedicate an area for the office. In my case, I've setup the back of the trailer that converts from seating space to a king size bed at night. When in the "office" mode, there is a table insert that serves as the desk between the bench style seating around the perimeter. See Figure 6.1 and Figure 6.2.

Figure 6.1 RV configured in "office" mode.

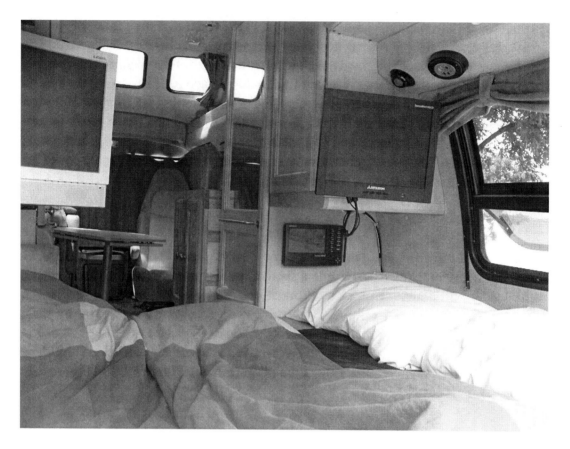

Figure 6.2 RV configured in "sleep" mode.

You will notice that I travel with a laptop computer to conserve space and energy consumption. I have added two LCD flat panel monitors that I connect to the laptop so that I have more computer "desktop" area to run multiple applications. I can have my e-mail open on one screen, Internet explorer for web browsing on another screen and my programming application running on the laptop's screen. I therefore reduce the number of times I have to minimize one program to view another, as is the case with having only one monitor. See Figure 6.3.

Figure 6.3 Multiple monitor setup.

My printer is tucked away at the end of the bench where seating is not available due to the A/V cabinet. I use a scanner to make electronic copies of important receipts, documents and manuals. In electronic format, I can send items via e-mail when need be, since I don't use a fax machine. The scanner has the ability to scan items directly to Adobe™ PDF file format for sending documents across the Internet. I use the scanner in conjunction with the computer as a copy machine. Also, since files and documents are stored on my computer, I don't need a lot of paper filing space and they get backed up for safe keeping. I store my scanner in the A/V cabinet when it's not in use.

The A/V cabinet is also where I keep my DVD player, external disk drive (for backups), CD/DVD disks, reference books, and other office supplies (see Figure 6.4). It has electrical outlets that are tied to shore power, as well as off the inverter. I can keep all my various electrical connections and transformers tucked out of site in the cabinet.

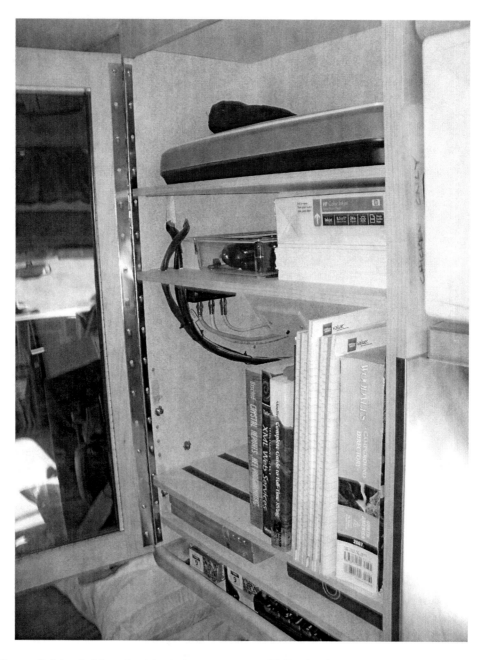

Figure 6.4 Audio/Visual cabinet stores scanner, DVD player, external disk drive, disks, books and office supplies.

Besides my power connections stored in the A/V cabinet, this is where all my cables are routed from the roof-mounted solar panels, Wi-Fi and television antennas, and the booster antenna for my cell phone. In my case, I drilled a hole through the fiberglass roof as an ingress point above the cabinet to feed in the cables. I added a rubber grommet and then used a waterproof roof patching compound to seal the hole.

All of my computer's external equipment is connected via USB connections. Because I have more connections than the three available on the laptop, I added a powered USB hub (as described in chapter 2) to increase my number of ports from 3 to 6.

The RoadTrek 190 came installed with two deep-cycle house batteries, inverter, and charging system (from shore power or engine alternator). I added two more auxiliary deep-cycle batteries for additional energy storage, an inverter, and a solar charger. I have two banks of solar cells, one for the RoadTrek batteries and one for the auxiliary batteries. See Figure 6.5, Figure 6.6, Figure 6.7, and Figure 6.8. I decided to go with two separate independent systems for redundancy, and because of a lack of room to install a switch gear to manage the battery charging functions (the batteries can not be charged from the solar panels AND the shore power/alternator at the same time). Alternately, I could have tied the four batteries in parallel, but since I installed high performance, maintenance-free AGM batteries for the auxiliary batteries, and the RoadTrek's house batteries were of the wet-cell design, that was not possible due to the recharging method being different between the two types. See the sidebar section in chapter 4 regarding **battery charging** for more details. The other motivating factor for separating the two battery banks was the logistics involved with running the thick gauge copper wiring between the batteries.

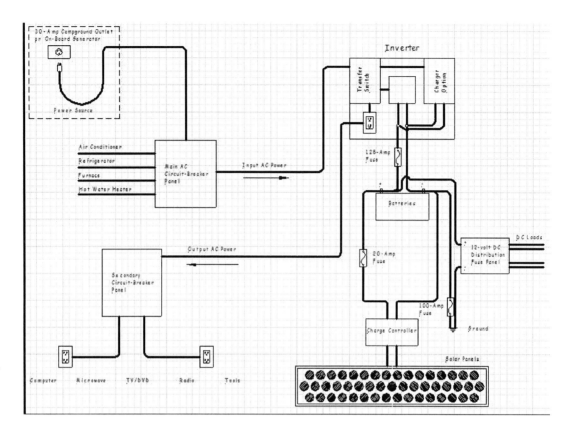

Figure 6.5 Electrical schematic for the RoadTrek 190 electrical storage and charging system.

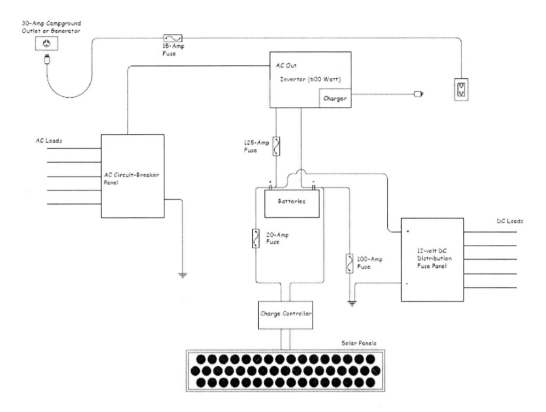

Figure 6.6 Electrical schematic for the auxiliary electrical storage and charging system.

194 The Complete Guide to RV Electrical, Computer, Solar and Communications Systems

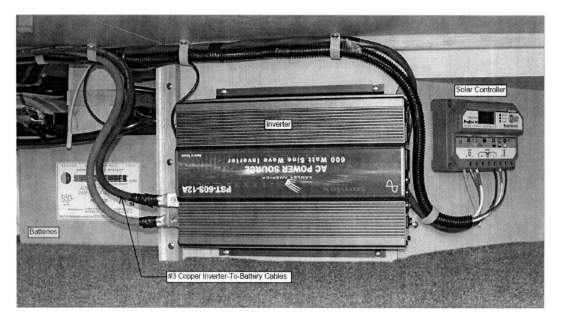

Figure 6.7 Bank A showing the RV batteries, inverter, and solar controller.

Figure 6.8 Solar collectors.

Testing

In designing the solar components and energy storage system (batteries), I built a solar test lab to determine what my needs would be. Since you've purchased this book, you can by-pass this step (unless you would like to try it), and just read the conclusions from my tests.

Initially, I made a rough estimate of what my power requirements would be for the equipment I planned to use including lighting and other miscellaneous electrical requirements. Table 6.1 shows how I arrived at my daily power usage measured in amp-hours. (The power for this equipment is supplied by the batteries and the batteries are charged by the solar cells when not operating on shore power.)

Item	Amps	Time Used	Amp-Hours Consumed
Laptop	3.2	4.0	12.8
15" LCD Monitor #1	2.5	4.0	10.0
17" LCD Monitor #2	3.0	4.0	12.0
Printer	1.0	0.5	0.5
Fan	1.0	2.0	2.0
3 Lights	3.6	3.0	10.8
Water pump	3.8	0.5	1.9
Misc.	2.0	2.0	4.0
		Total:	54.0

Table 6.1 Estimate of daily power usage.

Next, I researched the various components that would be required (solar panels, solar chargers, inverters, batteries, cables and fuses), and ended up purchasing two Kyocera model KC-65 solar panels rated at 65 watts each, two Concord SunExtender PVX-1040T 120 amp-hour AGM batteries, a Morningstar model ProStar PS-30 solar controller rated at 30 amps, a Samlex model PST-60S-12A 600-watt inverter, and various cables and fuses (See **Vendor Reference List**).

In the solar test lab, I put the system together and did a dry run to make sure everything was assemble properly. I used a Xantrex Trace™ model TM500A battery status monitor to obtain and record battery voltage, state-of-charge, amps being drawn, and the number of amp-hours consumed.

Solar panel test stands

Meter for measuring battery charge levels, voltage, amperes and amp-hours.

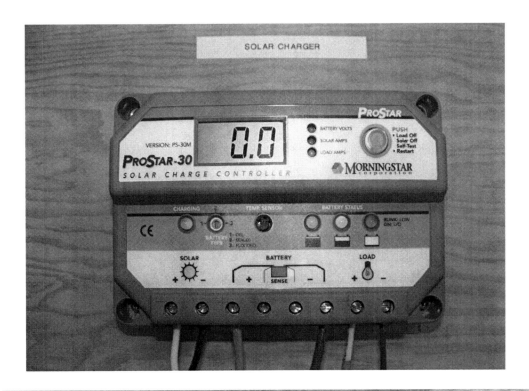

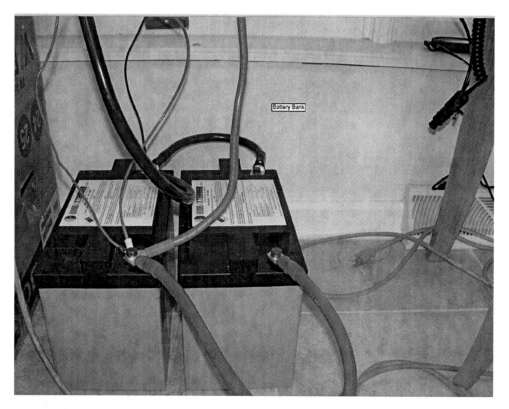

Computer lab workstation

During a month of testing[1] I simulated the actual environment by connecting the items listed in Table 6.1 to the inverter for AC loads and off the Load connection of the solar controller for DC loads.

There were three objectives I wanted to cover in these tests – 1) How long would the batteries hold out under my estimated power usage with no solar input?, 2) How long would it take me to fully recharge the batteries (assuming no load and clear skies) from a 50% state-of-discharge?, and 3) How long might I expect the batteries to hold out with a moderate amount of periodic sunshine?

First, I measured the discharge rate. With all loads in place, I recorded the amount of solar energy being received by the panels, number of amps drawn, battery voltage, and number of amp-hours consumed. I also measured the time it took for fully-charged batteries to get to a 50% state-of-charge both with and without solar input.

Next, I measured the recharge rate. With the batteries discharged to 50% capacity, I tested the recharge time by removing all loads and observing how long it took to get them back to 100% charge. I observed that in most trials, it took almost a full day with moderate sunshine to get the batteries back to 80%, another full day to get them back to 90% and yet another full day to get back to a 100% state-of-charge.

Conclusions

Based on the results, I concluded that I should be able to meet my daily power consumption with little or no solar input for 3.5 days.[2] I compared this to my theoretical calculation of 3.7 days.[3] Since this was close to my goal of having 4 days reserve, I concluded that the system components I selected in conjunction with the RoadTrek's battery capacity would meet my objectives.

Since the recharge time is quite lengthy when the batteries are discharged to a state-of-charge of 50%, I concluded it would be best to keep an eye on this level (either with a meter or by measuring the battery voltage), and not let the state-of-charge of the batteries drop much below 70% (12.3 volts). With the solar charging option, this is easier to maintain since the batteries are getting replenished whenever the sun is out.

---

[1] The tests were run at 43° N latitude, -88° W longitude during the month of March. Solar radiation for this area of the United States averages 400 kW/M^2, and there are roughly 4 hours of sun per day at that time of year. The batteries were kept in a 70° F climate, which yields 96% battery efficiency. Tests were done with panels at 45° and at 0° (flat).

[2] The test showed that 188.4 AH were used in discharging the batteries from 100% to a 50% state-of-charge. Using my estimated daily power usage of 54 AH, the batteries would last 188.4 AH / 54 AH / Day = 3.5 days.

[3] With the installed battery capacity of 400 AH and using a maximum discharge of 50%, the batteries would last (400 * 0.50) AH / 54 AH / Day = 3.7 days.

Appendix A – Vendor Reference List

This appendix contains a list of the vendors and web sites for the products and services that were mentioned in this book. The service column indicates the services or products which the vendor supplies, although vendors may offer more products and services than listed below. I've also included some extra companies that may be of interest for the RV enthusiast.

Vendor	Web Address	Service
AuthSMTP	www.authsmtp.com	Receiving e-mail anywhere
Camping World	www.campingworld.com	RV and camping supplies
Cannon Inc.	www.canon.com	Printers and ink supplies
Yagiantenna, Inc.	www.cellphonesolutions.com	Cell phone antennas
Charging Systems International	www.dualpro.com	Battery chargers
Custom Battery Cables	www.custombatterycables.com	Battery cables
DirecStar	www.direcstar.com	Satellite TV and Internet access
Don Rowe	www.donrowe.com	Solar components
Ergonomic Resources	www.ergonomicsmadeeasy.com	Wireless keyboards and mice
Fleeman Anderson & Bird	www.fab-corp.com	Wireless adapters and antennas
Hanmar Motor Corporation	www.roadtrek.com	RV manufacturer
Hewlett-Packard	www.hp.com	Scanners, printers and ink supplies
LinkSys	www.linksys.com	Wireless routers and networking
Memorex	www.memorex.com	Flash drives for data storage
Microsoft Corporation	www.microsoft.com	Windows and software
Monster Cable Products	www.monstercable.com	Surge protectors and cables
Morningstar Corporation	www.morningstar.com	Solar controllers
On-Line Solar, Inc.	www.mrsolar.com	Solar panels, cables
Net Zero	www.netzero.net	Dialup Internet access
P3 International	www.p3international.com	Watt meters
Radio Shack	www.radioshack.com	Electrical supplies
Samlex America	www.samlexamerica.com	Inverters
Satellites In Motion	www.satellitesinmotion.com	Mobile internet
Sports Imports LTD	www.sportsimportsltd.com	DC appliances
The Weather Channel	www.weather.com	Weather information
USB Gear	www.usbgear.com	USB devices
Technical Connections, Inc.	www.videomountstore.com	Flat panel mounting systems
Wal-Mart	www.walmart.com	Free overnight rest stop
West Marine	www.westmarine.com	Fuses and cables
Western Digital	www.westerndigital.com	Data storage
Wholesale Solar	www.wholesalesolar.com	Solar products
Wilson Electronics, Inc.	www.wilsonelectronics.com	Cellular antennas
Winegard	www.winegard.com	RV antenna systems
Woodall's	www.woodalls.com	Campground/Resort directory
Xantrex	www.xantrex.com	Monitoring meters

Appendix B - Definitions

This appendix contains the definitions for terms and abbreviations used throughout this book.

.bmp, .jpg, .gif - File name extensions associated with files created by graphics programs. These extensions are usually associated with photographs.

.doc - The file name extension for files created by Microsoft® Word.

.inf - The file name extension for files that contain device information or scripts to control hardware operations.

.pdf - The file name extension for files created by Adobe® Acrobat.

.pst - The file name extension for data files created by Microsoft® Outlook.

AC - In electrical circuits, this refers to Alternating Current.

AGM – Absorbed Glass Mat is a material used in the construction of deep-cycle batteries.

AH - Battery capacity measured in Amp Hours.

Broadband - A high-speed connection. Broadband connections are typically 256 kilobytes per second (KBps) or faster. Broadband includes DSL and cable modem service.

Boot - The process of starting or resetting a computer. When first turned on (cold boot) or reset (warm boot), the computer runs the software that loads and starts the computer's operating system, which prepares it for use.

Byte - A unit of data that typically holds a single character, such as a letter, a digit, or a punctuation mark. Some single characters can take up more than one byte.

CO - In the telephone industry it means Central Office which is a building that houses, among other things, the call switching equipment to connect calls.

CRT - A type of video monitor that uses a vacuum tube containing an electron gun (a source of electrons) known as a Cathode Ray Tube to display an image on a florescent screen.

DC - In electrical circuits, this refers to Direct Current.

Desktop - The on-screen work area on which windows, icons, menus, and dialog boxes appear.

DHCP - A service protocol that offers dynamic leased configuration of host IP addresses and distributes other configuration parameters to network clients. Dynamic Host Configuration Protocol provides network configuration, prevents address conflicts, and helps conserve the use of client IP addresses on the network.

The DHCP server maintains centralized management of IP addresses that are used on the network. DHCP-supported clients can then request and obtain lease of an IP address from a DHCP server as part of their network boot process.

DNS - The Domain Name System is a hierarchical, distributed database that contains mappings of DNS domain names to various types of data, such as IP addresses. DNS enables the location of computers and services by user-friendly names, and it also enables the discovery of other information stored in the database.

Domain - A group of computers that are part of a network and share a common directory database. A domain is administered as a unit with common rules and procedures. Each domain has a unique name and is made up of user groups, computer groups, printer groups and other resources.

DSL - Digital Subscriber Line is a type of high-speed Internet connection using standard telephone wires. This is also referred to as a broadband connection.

Firewall - A combination of hardware and software that provides a security system, usually to prevent unauthorized access from outside to an internal network or intranet. A firewall prevents direct communication between a network and external computers by routing communication through a proxy server outside of the network. The proxy server determines whether it is safe to let a file pass through to the network.

Flash Drive - A small external storage drive for holding programs, files and data. It is often used as a backup device for data or for transporting data from one computer to another. It connects via a USB port. It is also sometimes referred to as a thumb drive or memory stick.

Gateway - A device connected to multiple networks capable of routing or delivering IP packets (data) between them. A gateway translates between different transport protocols or data formats and is generally added to a network primarily for its translation ability.

GB - A unit of data used to describe a device's memory capacity. One gigabyte is approximately one billion bytes.

GHz - A term that is used in electronics to describe the frequency (cycles per second) of a device or radio signal. One Gigahertz is approximately one billion cycles per second. For example, a 3 GHz processor indicates that a computer processes instructions at a speed of three billion cycles per second.

GPS - The Global Positioning System uses satellite triangulation to locate a position on the Earth's surface.

Hard Drive - A hard drive, also called a hard disk, contains one or more inflexible platters coated with material in which data can be recorded magnetically with read/write heads. The hard drive exists in a sealed case that protects it and allows the head to fly 10 millionths to 25 millionths of an inch above the surface of a platter. Data can both be stored and accessed much more quickly than on a floppy disk.

Hotspot - A hotspot is a term used to describe a location such as a library, business, or campsite that has a wireless connection available for connecting to the Internet. A hotspot will often be described as having Wi-Fi service.

HTTP - Hypertext Transfer Protocol is the protocol used to transfer information on the World Wide Web. An http address (one kind of uniform resource locator [URL]) takes the form of http://www.microsoft.com.

Hub - A common connection point for devices on a network. Typically used to connect segments of a local area network (LAN), a hub contains multiple ports. When data arrives at one port, it is copied to the other ports so that all segments of the LAN can see the data.

IEEE - The Institute of Electrical and Electronics Engineers is a professional organization for the advancement of technology related to electricity.

Intranet - A network within an organization that uses Internet technologies and protocols, but is available only to certain people, such as employees of a company. An intranet is also called a private network.

Internet - The worldwide network of computers. If you have access to the Internet, you can retrieve information from millions of sources, including schools, governments, businesses, and individuals.

Internet Address - An address for a resource on the Internet that is used by web browsers to locate Internet resources. An Internet address typically starts with a protocol name, followed by the name of the organization that maintains the site and the suffix identifies the kind of organization it is. For example, the address *http://www.yale.edu* provides the following information:
- *http*: This web server uses the Hypertext Transfer Protocol.
- *www*: This site is on the World Wide Web.
- *edu*: This is an educational institution.

An Internet address is also called a Uniform Resource Locator (URL).

Inverter - A power inverter changes DC power from a battery into conventional AC power that can be used to operate electric lights, kitchen appliances, microwaves, power tools, TVs, radios, computers, etc.

IP - The Internet Protocol is a routable protocol in the TCP/IP protocol suite that is responsible for IP addressing, routing, and the fragmentation and reassembly of IP packets.

IP Address - A 32-bit address used to identify a node (e.g. a computer, a printer, a server, etc.) on an IP network. Each node on the IP network must be assigned a unique IP address, which is made up of the network ID, plus a unique host ID. This address is typically represented with the decimal value of each octet separated by a period (e.g.192.172.7.3).

ISP - Internet Service Provider. A company that provides individuals or companies access to the Internet and the World Wide Web. An ISP provides a telephone number, a user name, a password, and other connection information so users can connect their computers to the ISP's computers. An ISP typically charges a monthly or hourly connection fee.

LAN - A communications network (local area network) connecting a group of computers, printers, and other devices located within a relatively limited area (for example, a building). A LAN allows any connected device to interact with any other on the network.

LCD - A Liquid Crystal Display is a type of display used in laptop computers and flat panel monitors.

LVD - Low Voltage Disconnects are used in solar charge controllers to reduce damage to batteries by avoiding deep discharge. When the battery reaches a preset voltage (usually around 12.1 volts), an LVD disconnects the load attached to the battery.

MB - A unit of data used to describe a device's memory capacity. One megabyte is approximately one million bytes.

Media - Any fixed or removable objects that store computer data. Examples include hard disks, floppy disks, tapes, and compact discs.

MHz - A term that is used in electronics to describe the frequency (cycles per second) of a device or radio signal. One Megahertz is approximately one million cycles per second. For example, a 500 MHz indicates that a computer processes instructions at a speed of five hundred million cycles per second..

Modem – Short for Modulator/Demodulator, it is a device that allows computer information to be transmitted and received over a telephone line. The transmitting modem translates digital computer data into analog signals that can be carried over a phone line. The receiving modem translates the analog signals back to digital form.

Multiple Boot - A computer configuration that runs two or more operating systems. It can also be referred to as a dual boot.

Network - A group of computers and other devices, such as printers and scanners, connected by a communications link, enabling all the devices to interact with each other. Networks can be small or large, permanently connected through wires or cables, or temporarily connected through phone lines or wireless transmissions. The largest network is the Internet, which is a worldwide group of networks.

Network Adapter - A device that connects your computer to a network. This device is sometimes called an adapter card or network interface card (NIC).

Omni Directional - Refers to an antenna, for example, that can receive signals from all directions.

PC - Personal computer.

Plug and Play - A set of specifications developed by Intel that allows a computer to automatically detect and configure a device and install the appropriate device drivers.

POP3 - Post Office Protocol 3. A popular protocol used for receiving e-mail messages. This protocol is often used by ISPs. POP3 servers allow access to a single Inbox.

Port - A connection point on your computer where you can connect devices that pass data into and out of a computer. For example, a printer is typically connected to a parallel port (also called an LPT port), and a modem is typically connected to a serial port (also called a COM port).

Profiles – A profile is a set of instructions and settings for a computer user or program. For example, when you logon to your computer, your profile is accessed to display your desktop appearance (such as color, background image, icons, screensaver, etc.), items on the Start menu and other personal settings.

Properties - A characteristic or parameter of a class of objects or devices. For example, properties of Microsoft Word files include Size, Created, and Characters. Properties are found by right-clicking your mouse on things like icons, folders and files.

Protocol - A set of rules and conventions for sending information over a network. These rules govern the content, format, timing, sequencing, and error control of messages exchanged among network devices.

PWM - Pulse Width Modulation is a feature of battery chargers and solar controllers. In PWM, the controller or charger senses tiny voltage drops in the battery and sends very short charging cycles (pulses) to the battery. This may occur several hundred times per minute. It is called "pulse width" because the width of the pulses may vary from a few microseconds to several seconds.

RAM - Random Access Memory is computer memory that can be read from or written to by a computer or other devices. Information stored in RAM is lost when the computer is turned off.

Registry - A database repository for information about a computer's configuration. The registry contains information that Windows continually references during operation, such as:

- Profiles for each user.
- The programs installed on the computer and the types of documents each can create.
- Property settings for folders and program icons.
- The hardware that exists on the system.
- The ports which are being used.

Right click - To position the mouse over an object, and then press and release the secondary (right) mouse button. Right-clicking opens a shortcut menu that contains useful commands, which change depending on where you click.

ROM - An acronym for Read Only Memory, a semiconductor circuit into which code or data is permanently installed by the manufacturing process. ROM contains instructions or data that can be read but not modified.

Router - Hardware that helps LANs and WANs achieve interoperability and connectivity, and can link LANs that have different network topologies (such as Ethernet and Token Ring). Routers match packet headers to a LAN segment and choose the best path for the packet, optimizing network performance.

Server - In general, a computer that provides shared resources to network users.

SLI - In battery terminology this is an acronym for Starting, Lights and Ignition. Automotive batteries are often referred to as SLI batteries.

SMTP - Simple Mail Transfer Protocol is a member of the TCP/IP suite of protocols that governs the exchange of electronic mail between message transfer agents.

SSL - Secure Sockets Layer is an open standard for establishing a secure communications channel to prevent the interception of critical information, such as credit card numbers. Primarily, it enables secure electronic financial transactions on the World Wide Web, although it is designed to work on other Internet services as well. You are accessing a secure web site when the Internet address takes the form of, for example **https**://www.paypal.com. The https: portion of the address means that this web server uses the secure Hypertext Transfer Protocol.

TCP/IP - An acronym for Transmission Control Protocol/Internet Protocol. A set of networking protocols widely used on the Internet that provides communications across interconnected networks of computers with diverse hardware architectures and various operating systems. TCP/IP includes standards for how computers communicate and conventions for connecting networks and routing traffic.

URL - An address that uniquely identifies a location on the Internet. A URL for a World Wide Web site is preceded with http://, as in the URL http://www.microsoft.com. A URL can contain more detail, such as the name of a page of hypertext, usually identified by the file name extension .html or .htm.

USB - The Universal Serial Bus is a standard that supports Plug and Play installation. Using USB, you can connect and disconnect devices without shutting down or restarting your computer. You can use a single USB port to connect up to 127 peripheral devices, including speakers, telephones, CD-ROM drives, joysticks, tape drives, keyboards, mice, scanners, and cameras.

Virtual Memory - Temporary storage used by a computer to run programs that need more memory than the computer has. For example, programs could have access to 4 gigabytes of virtual memory on a computer's hard drive, even if the computer has only 32 megabytes of RAM. The program data that does not currently fit in the computer's memory is saved into paging files that are stored on the computer's hard drive.

WAN - A Wide Area Network connects geographically separated computers, printers, and other devices. A WAN allows any connected device to interact with any other on the network.

WAP - Used in wireless networks, the Wireless Access Point does the routing between the local network and the public network. (This is also referred to as the gateway.)

Wi-Fi - Wireless Fidelity is the common name for a popular wireless technology used in home and business networks.

Wireless - Short for wireless communication between a computer and another computer or device without wires.

Wizard - A special application that assists you with installing programs and computer hardware. It presents step-by-step instructions and often does much of the setup on its own using the common default settings, but allows you to override them if desired.

Workgroup - A simple grouping of computers, intended only to help users find such things as printers and shared folders within that group. Workgroups in Windows do not offer the centralized user accounts and authentication offered by domains.

WWW - Acronym for the World Wide Web. It's a system for exploring the Internet by using hyperlinks. When you use a web browser, the web appears as a collection of text, pictures, sounds, and digital movies.

WZC - The Windows Zero Configuration service is the wireless connection manager installed with Windows. It assists in making wireless networking connections.

Appendix C – Helpful Links

Area Code Map (http://www.nanpa.com/area_code_maps/ac_map_static.html)
Aviation Weather (http://adds.aviationweather.gov/)
Boondocking (Free Camping) (http://www.boondocking.org/)
BOONDOCKING!!! (http://www.angelfire.com/oh/Boondocking/)
Local and National Yellow Pages (http://www.smartpages.com/)
More Boondocking (http://www.rvhometown.com/HTML/Articles/Boondocking.htm)
NANPA Area Code Map (http://www.nanpa.com/area_code_maps/ac_map_static.html)
Recipes (http://www.cooks.com/)
RV Bookstore (http://rvbookstore.com/)
RVTraderOnline (http://www.rvtraderonline.com/)
The official U.S. time clock (http://www.time.gov)
The Evening & Morning Star (http://www.johnpratt.com/items/astronomy/eve_morn.html)
Wal-Mart Camping and Other Free Places (http://www.legendsofamerica.com/RV-Walmart.html)
Weather (http://www.weather.com)
Weather Radios (http://www.weatherradiostore.com/)
Wi-Fi Hot Spots (http://www.jiwire.com/search-hotspot-locations.htm)
Zip Codes (http://www.getzips.com/city.htm)

Index

AC System 66, 69
Amperage Draw 67, 74
Antenna 4, 28-32, 115, 125, 191, 208
Batteries 14, 74, 76-83, 90, 94-98, 103, 105, 108-114, 195, 200, 201
   See also Battery Charging
Battery Charging 83, 98, 108, 109, 114, 191
Boondocking 35, 187
Cable TV 123, 126
Cellular 115
Charge Controller 78, 99, 105, 106, 114
Circuit Analyzer 70, 71
Circuit Breaker 4, 67, 68, 105, 113
Computer 7, 8
Computer Backup 37-49, 55-63, 128
CRT 8
DC System 73, 74
DHCP 28
Dial-up Connections 115, 122, 128, 129, 134, 138, 140, 141
DirecTV 126
   See also Satellite TV
Dish Network 126
   See also Satellite TV
DNS 28, 120
Domain 145, 185
DSL 27, 123, 129
Electrical
   See AC System, DC System
Electrical Adapters 69, 70
Electrical Outlets 69, 70, 74, 83, 189
Electrical Schematic 2-5, 122-124, 192, 193
E-Mail 8, 34, 115, 119, 122, 143-185
Flash Drive 38, 58, 128
Fuse panel 4, 73
Fuses 105-107, 114
Generator 71, 74, 78, 87, 98, 105, 112, 113, 187

GPS (Global Positioning System) 1, 206
High Speed Internet 17, 23, 33, 122, 123, 125
Hot Spots 37, 124
Internet 7, 16-29, 32, 33, 37, 115, 117-125, 129
   See also High Speed Internet; World Wide Web
Inverter 4, 71, 74, 78, 83, 86-89, 107, 108, 111-114
IP Address 28, 120
Keyboards 11
Laptops 7, 9, 16, 76
LCD 1, 8, 9, 188
Meters 71, 90, 93, 105
Mice 11
Microsoft Outlook 144
   See also Outlook Data File; Outlook Profiles
Modem 73, 139, 140
Monitors 8-10, 87, 89
Multimeter *See* Meters
Outlook Data File 152, 157-162, 168
Outlook Profiles 151, 154, 171
Phones 115-117
   See also Cellular
Planning 1, 2, 67
Power Requirements 3, 74-76, 96, 100, 195
Printing 34
RJ-11 Jack 73, 128, 140
RV Floor Plans 2
RV Office 34-36, 66, 115, 187, 189
Safety 6, 83, 94, 106
Satellite Dish 4, 125-128
Satellite TV *See* Satellite Dish
Scanning 34, 36
Solar **98**
State of Charge 80, 90, 92, 105, 106, 108, 109, 195, 200, 201
Surge protector 73
System Ground 114

Transfer Switch 111, 112
URL (Uniform Resource Locator) 120, 121
USB (Universal Serial Bus) 221
   See also USB Ports; USB Connections; USB Hubs
USB Connections 9, 11, 13, 41, 128, 191
USB Hubs 13-15, 191
USB Ports 14
Vendors 202
Video Card 9
Voltmeter 90
Weight Capacity 6
Wi-Fi 7, 16, 23, 26-29, 32, 33, 37, 124, 125, 191
   See also Hot Spots
Windows Zero Configuration 16, 20
Wire Gage 4, 5
Wireless 11, 14, 16, 17, 22, 23, 26, 27, 29, 33, 124, 125
   See also Wireless Networking; Wireless Access Point
Wireless Access Point 16, 27, 29, 30, 32
Wireless Networking 16, 30, 125
Wiring Diagrams 111-113
Woodall's Campground Directory 124
World Wide Web 121